A USER'S GUIDE TO THE AGE OF TECH

Electronic Mediations

Series Editors
N. Katherine Hayles, Peter Krapp, Rita Raley, and Samuel Weber
Founding Editor: Mark Poster

65 *A User's Guide to the Age of Tech*
Grant Wythoff

64 *The Little Database: A Poetics of Media Formats*
Daniel Snelson

63 *Interactive Cinema: The Ambiguous Ethics of Media Participation*
Marina Hassapopoulou

62 *The Digital and Its Discontents*
Aden Evens

61 *The Digitally Disposed: Racial Capitalism and the Informatics of Value*
Seb Franklin

60 *Radical Secrecy: The Ends of Transparency in Datafied America*
Clare Birchall

59 *Perpetual Motion: Dance, Digital Cultures, and the Common*
Harmony Bench

58 *Playing Nature: Ecology in Video Games*
Alenda Y. Chang

57 *Sensations of History: Animation and New Media Art*
James J. Hodge

56 *Internet Daemons: Digital Communications Possessed*
Fenwick McKelvey

55 *What Is Information?*
Peter Janich, Eric Hayot, and Lea Pao

54 *Deconstruction Machines: Writing in the Age of Cyberwar*
Justin Joque

53 *Metagaming: Playing, Competing, Spectating, Cheating, Trading, Making, and Breaking Videogames*
Stephanie Boluk and Patrick LeMieux

52 *The Perversity of Things: Hugo Gernsback on Media, Tinkering, and Scientifiction*
Hugo Gernsback
Edited by Grant Wythoff

51 *The Participatory Condition in the Digital Age*
Darin Barney, Gabriella Coleman, Christine Ross, Jonathan Sterne, and Tamar Tembeck, Editors

50 *Mixed Realism: Videogames and the Violence of Fiction*
Timothy J. Welsh

49 *Program Earth: Environmental Sensing Technology and the Making of a Computational Planet*
Jennifer Gabrys

48 *On the Existence of Digital Objects*
Yuk Hui

47 *How to Talk about Videogames*
Ian Bogost

46 *A Geology of Media*
Jussi Parikka

45 *World Projects: Global Information before World War I*
Markus Krajewski

44 *Reading Writing Interfaces: From the Digital to the Bookbound*
Lori Emerson

43 *Nauman Reiterated*
Janet Kraynak

42 *Comparative Textual Media: Transforming the Humanities in the Postprint Era*
N. Katherine Hayles and Jessica Pressman, Editors

41 *Off the Network: Disrupting the Digital World*
Ulises Ali Mejias

40 *Summa Technologiae*
Stanisław Lem

39 *Digital Memory and the Archive*
Wolfgang Ernst

38 *How to Do Things with Videogames*
Ian Bogost

(continued on page 153)

A USER'S GUIDE TO THE AGE OF TECH

GRANT WYTHOFF

Electronic Mediations 65
University of Minnesota Press
MINNEAPOLIS | LONDON

Portions of chapter 1 are adapted from "Pocket Wireless and the Shape of Media to Come, 1899–1922," *Grey Room* 51 (2013): 40–63, https://doi.org/10.1162/GREY_a_00106. Portions of chapter 1 are adapted from "Ensuring Minimal Computing Serves Maximal Connection," *Digital Humanities Quarterly* 16, no. 2 (2022). Portions of chapter 2 are adapted from "Extended Technique: New Scholarship on the Uses of Media," *Amodern* 9 (2020).

Copyright 2025 by the Regents of the University of Minnesota

All rights reserved. No part of this publication may be reproduced, stored in a retrieval system, utilized for purposes of training artificial intelligence technologies, or transmitted in any form or by any means, electronic, mechanical, photocopying, recording, or otherwise, without the prior written permission of the publisher.

Published by the University of Minnesota Press
111 Third Avenue South, Suite 290
Minneapolis, MN 55401-2520
http://www.upress.umn.edu

ISBN 978-1-5179-1876-7 (hc)
ISBN 978-1-5179-1877-4 (pb)

A Cataloging-in-Publication record for this book is available from the Library of Congress.

Printed in the United States of America on acid-free paper

The University of Minnesota is an equal-opportunity educator and employer.

for Asa

CONTENTS

Preface: Technique in the Age of Tech ix

A primer on *tékhnē, technics, technique,* and associated portraits of the way we do things.

1. Sleight of Hand 1

Fashionable urbanites begin carrying wireless telegraphs in their pockets and wonder what to say to each other, 1919. Efforts to remain connected during social distancing measures show the fraying seams of that technoutopian vision, 2020. An introduction to the book's main themes.

2. The Way We Do Things 19

In war-torn France, Marcel Mauss struggles to complete a treatise on the philosophy of technique, 1941. Virtuoso smartphone gestures flicker on the subway, 2019. Applied scientists propose a humanist technology criticism, 1850s. A report from the high theory of technique.

3. *Technically Speaking* **43**

Sailors, mechanics, and other technicians invent new words for tools and techniques, 1840s. Lexicographers hunt for specimens of those words in use, 1888. Linguists describe the meaning of a word as the company it keeps, 1957. An excavation of the low theory of technicians.

4. *The Custody of Automatism* **71**

Community technologists design intentional, cooperative infrastructures by building local consensus, 2023. Smartphone addicts miss their preinternet brains, 2014. A case for conscientious technique.

Epilogue: Reclaiming Technique **95**

A diagnosis concerning technique as the fossil fuel of AI, and two possible futures for refusal.

Acknowledgments   101
Notes   105
Index   143

Naomi Elena Ramírez, *Choreography for Smartphone Gestures* (score for performance), giclée with pencil on single matte Mylar. 3 mil, 30 × 60 in.

PREFACE

Technique in the Age of Tech

THIS MORNING WHILE MAKING BREAKFAST, I SPENT about a minute using an app to decide whether to drive or take the bus to a doctor's appointment. That brief interaction relied on a device concocted from an alchemical mixture of rare earth elements sourced from around the planet and a network of satellites orbiting said planet in perfect synchronization. It summoned millimeter-size radio frequency waves precisely calibrated according to how quickly they're absorbed by a gram of tissue in my hand, and machine learning algorithms capable of generating predictions that are unexplainable even by their creators. Every day, I casually use systems of inconceivable complexity. This process bridged scales of magnitude from the molecular to the planetary, but to me, it was second nature. A few practiced taps of the screen, and I was ready to go.

This book explores how we—users of tools nearly indistinguishable from magic—have understood the presence of those tools in our daily lives, and how we have adapted them toward unexpected ends. Each of us develops our own techniques to wield tools as complex as the smartphone, concepts as abstruse as electromagnetism. In the opening quarter of the twenty-first century, those techniques have undergone unprecedented changes spanning the mass adoption of the internet and the mass adoption of AI. The pages that follow take stock of the ways users past and present negotiate such changes.

So where to begin? It's easy to describe a device like the smartphone empirically, but the use of that device is an ephemeral performance, culturally contextual, and unique for everyone. One classic starting point comes from the concepts of *tékhnē* and *epistēmē:* knowing how to do something, and knowing that something is true. The distinction is ancient. Modern life, however, troubles the divide between *tékhnē*'s craft, skill, and habit on the one hand and *epistēmē*'s ideas, concepts, and assertions on the other. Whether we're taking a shortcut around traffic or setting a timer to make a perfectly soft-boiled egg, there's knowledge of the world bound up in the way we do things. In the case of information technologies—devices, infrastructures, media, data—the distinction between knowing-how and knowing-that becomes even fuzzier. Of course these technologies involve specific procedures for their use. But because they also encode forms of knowledge, it's a challenge to delineate the beginnings and ends of tool, technique, and understanding.

Modern theories of *tékhnē* provide countless ways of studying the relationship between tools and their use, and a number of them are woven throughout this book. Canonical approaches range from Michel Foucault's techniques of the self to Marcel Mauss's body techniques, the latter of which will receive a detailed treatment here.[1] There have been labor theories like Ruth Schwartz Cowan's work process and Lucy Suchman's plans and situated actions.[2] Descriptions of social media tactics come from Tero Karppi and Trine Syvertsen on refusal and André Brock on Black digital practice.[3] Approaches to the gender of technique can be found in Paul Preciado's notion of prosthetic masculinity and Zoë Sofia's work on container technologies.[4] We have rich anthropological accounts of craft, affordances, and entanglement from Richard Sennett, Tim Ingold, and Ian Hodder.[5]

Today, most media theorists associate *tékhnē* with the German concept of cultural techniques *(Kulturtechniken)*. Scholars like Sybille Krämer, Bernhard Siegert, Cornelia Vismann, and Geoffrey Winthrop-Young describe cultural techniques as ensembles of gestures, procedures, tools, and infrastructures that operate together to generate distinctions like "inside/outside, pure/impure, sacred/profane, female/male, human/animal, speech/absence of speech, signal/noise, and so on."[6] We could add to this list of academic concepts the professional practices developed by contemporary fields like human-computer interaction and user experience design. Our abbreviated list could be further enriched by sources from global antiquity: the second millennium BCE Sanskrit production of *mantras* (thought instruments), the fourth-century BCE concept of *wu-wei* (effortless action) in Daoist thought, or the ninth-century CE *Book of Ingenious Devices,* a Persian manual of procedurally directed, animate machines based on Byzantine, Chinese, and Indian sources.[7]

There are a thousand possible routes across concepts like these, a hundred different depths at which to plumb the lines of influence: *Kulturtechnik* is informed by the "operational sequence" coined by André Leroi-Gourhan, who first got the idea from Mauss's body techniques, who drew direct inspiration from Augustus Pitt-Rivers, &c. &c.[8] While this book contains just one of many possible routes across these concepts and through their influences, it does propose a common axis for comparing them.

All the sources listed above rely on different terminologies even though their ideas rhyme: *tékhnē,* interaction, affordance, *Technik,* know-how. My humble proposal is to reset our terminological tangle for machines and media theories by returning to the common language of *technique.* In the mid-twentieth century, some attempted to use an English-language neologism—*technics*—to capture the broad range of meanings bound up in these overlapping approaches to complex technical systems and the ways they're put to use.[9] But the use of *technics* by thinkers like Lewis Mumford, Thorstein Veblen, and Langdon Winner never caught on widely, save for a Japanese brand of hi-fi stereos and turntables. Throughout this book, I have sought to enrich the everyday sense of *technique* rather than rely on the foundational discourse on *tékhnē* or the neologism *technics.* While other keywords swirl around us, shift senses, and fall into disuse, the common word *technique* steadily hums along, sturdy enough to be infused with new meaning. This study explores the ideas, arguments, and fictions that are embedded in the execution of our everyday habits. It is grounded, therefore, in the often mundane procedures and skills best captured by the word *technique*.

My understanding of *technique* will be laid out and refined over the course of the book. For now, I will say that there are three main reasons why it became my privileged keyword in the process of wrangling with these ideas. First, *tékhnē* in the 2020s is weird. The theory's earliest models were designed for hand tools used by artisans who honed their craft over a lifetime of practice. Much later, the models were retrofitted for the harsh factories and machines of the industrial age, when the tool made use of the alienated worker.[10] Today, the tools we use—smartphones, social media, AI—are constantly trying to influence how, when, and why we use them, recording and learning from our every decision in the process. Attempting to build new theories and politics suited for the analysis of such distributed, complex systems often directs our focus to the macro level of big data, server farms, and algorithms. But the common language sense of *technique* as the way we do things grounds the political implications of those systems in the concrete experiences of everyday life. In this book, body techniques are the central locus around which the full stack of information technology operates in the twenty-first century: digital devices, user interfaces, algorithms, network infrastructures, data centers, aesthetic forms, power.

Second, embracing the functional connotations of *technique* means hewing closely to the praxis of workers, technicians, and users. The translators of Gilbert Simondon's *On the Mode of Existence of Technical Objects* explain that the word *technics* is "differentiated in English from *technique* insofar as the latter refers to the almost ineffably practical and particular application of technics to a given concrete task."[11] From the vantage point of this book, that is a feature of the word, not a bug. Starting with the relatively narrow English *technique* as a base and infusing it with new senses (many of which are already found, as we will see, in the cognates of *technique* in countless other languages) is a means of grounding theory in practice. In the chapters that follow, I hold academic concepts in equal regard with the habits, beliefs, and folk theories of users. My goal is to diversify the forms of expertise that count in conversations on technology, privacy, and the public. At a time when data brokers seek to know us better than we know ourselves, this book takes stock of the incisive things that we users know about tech.

Finally, *technique* may not have the same conceptual specificity that *technics* does for the study of information technologies. But the context-dependent nature of the word *technique* as it's used in countless other areas of practice—music, sports, cooking, magic—make it a deep repository of analogies for our experience as users of digital devices. I draw throughout this book on the language of technique in cooking and music to cast new perspectives on the use of digital devices, which too often feels without quality or skill. No one tells us how to use a smartphone; it's just supposed to work. But in other domains, it is possible to have good or bad technique, depending on training (or lack thereof). Seen in a comparative light, the practical advantages of *technique*'s seeming one dimensionality unfolds as we consider its meaning in the kitchen or on the stage, or trace what happens when techniques are imported from one domain of practice into another.

This book is grounded in concepts from media theory, digital humanities, and science and technology studies. My sources include tech critics, community technology activists, late nineteenth-century and early twentieth-century newspapers, and contemporary novels. To animate these concepts and sources, I consider the embedded use of various tools like plows, large language models, knots, dictionaries, ladders, and radios. I've attempted to braid these diverse influences together using a writing style that finds common ground between them. In doing so, my goal has been to invite many kinds of readers to draw their own connections across unexpected contexts. Perhaps they will walk away sharing my conviction that the way we do things is always more important, more interesting, than the tools we use to do them.

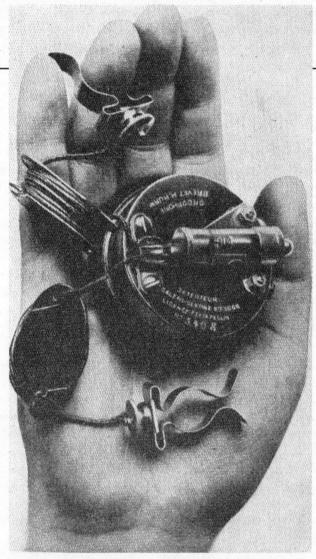

Horace Hurm's Ondophone, a portable wireless telegraph receiver sold in Paris. "Vest-Pocket Wireless Receiving Instrument," *Electrical Review and Western Electrician*, April 11, 1914, 745. Photographed by Digital Imaging Studio, Princeton University Libraries.

1

Sleight of Hand

A CENTURY AGO, AMID ANOTHER PANDEMIC AND NOT far from where I live, reporters flocked to the home of a man who claimed he could place a wireless telephone call. They were met by William MacFarlane, a Philadelphia experimenter wearing headphones, wrapped in wires, and holding three pieces of black stovepipe cylinder above his head. "I am working on a new theory in electricity, and in wireless," he announced. "I think the accepted principle of the waves is only part of the story. There is something else. It is that something else that I am utilizing."[1] Not one for lingering on details, MacFarlane instead staged elaborate scenes of communication for the assembled reporters. Packed in the back of his speeding car, they wore headphones and listened in as MacFarlane called his wife, who waited at the garage: "We are going to run down the road. Can you hear me?" From several hundred yards away, her reply came: "Yes, perfectly. Where are you?" The wind rushed through the stovepipe aërial—an early term for antenna—as everyone wondered at a twelve-pound black box between MacFarlane's feet; MacFarlane refused to answer questions about its contents.[2] Despite the box's size and weight, the reporters called the gadget a pocket wireless.

When they returned to the garage, MacFarland put on the air of a general and demonstrated how to command troops from afar. The reporters watched as a soldier marched across a distant field—perhaps a neighbor donning the uniform he had boxed up when the Great War came to an end seven months earlier. The rifle slung over his shoulder doubled as an antenna to collect the waves conveying his general's instructions. Though the soldier remained out of earshot, his march changed direction according to the commands MacFarlane spoke into the pocket wireless microphone. "If this could have been ready for use in the war," he told the reporters, "think of the value it would have had. A whole regiment equipt with the telephone receivers, with only their rifles as aërials, could advance a mile and each would be instantly in touch with the commanding officer. No runners would be needed. There could be no such thing as a 'lost battalion.'"

In 1919, most pocket wireless experimenters had only succeeded in sending Morse code messages. Beaming the human voice through a handheld device was an impressive accomplishment, especially given the physics.[3] The MacFarlanes' contraption would have operated at around 1,500 kilohertz, similar to AM radio today: lumbering, 650-foot-long waves so fragile that they shattered against buildings and lost their form, waves that canceled out any other signal in the vicinity trying to reach its destination. (Our smartphones, by contrast, operate at around 11 gigahertz, which means that the delicately interlaced packets of information carrying our emails, notifications, and texts ride radio waves little more than an inch long.[4]) But this prototype was enough to impress the audience. So carried away was one reporter by the possibility of hearing voices from the void—of holding a conversation with a loved one amid the distance and isolation of war and the pandemic—that the reporter wrote, "pocket wireless, that elusive objective of science, is only a step away."[5]

The MacFarlanes' pocket wireless was just one of countless others that amateur experimenters around the world were building from available household materials.[6] Activity was especially frenetic in the United States, where the federal government repealed a wartime ban on amateur wireless just two months before this Philadelphia demonstration.[7] Curtain rods and bicycles joined rifles to form aërials while metal-heeled shoes, walking canes, and—counterintuitively—wired telegraph poles served to ground the user, whose entire body formed a resonant electric circuit. Some stood on manhole covers, angling a top-hat antenna to take in the day's stock quotes. Others held umbrellas aloft, wrapping the wires of their vaguely stethoscopic leads around a fire hydrant to listen for a friend. The technique depended on the users, the materials, and what they hoped to hear.

Pocket wireless, doomed to be only a minor figure in media history, was never fully adopted and was quickly passed over in favor of devices and modes of communication that proved more practical. But it constituted a pattern of expectation, a dream of ubiquitous connection, presence, and belonging that held sway for years over the reception of wireless telegraphy: the technology that would become radio.[8] The concept of this medium was a product of its historical moment, with the war and the pandemic both animating countless excited visions. This was especially the case in Philadelphia, an epicenter of what was at the time called the Spanish flu. With all of the country's medical supplies and personnel going toward the war in Europe, the city's residents were on their own. They formed local aid societies to provide medical care and keep track of case counts. At the pandemic's peak, one of those volunteer groups declared that "the death toll for one day in Philadelphia alone was greater than the whole death toll from France for the whole American Army in one day."[9] Walking around the city, the MacFarlanes would have passed signs warning "Spit Spreads Death." With the mortuaries overwhelmed, they would have seen pieces of cloth hanging from the doors of crowded tenement buildings: a sign left by the multiple despairing families living there that a dead body was inside, with nowhere to go.[10]

Pocket wireless would have promised the circulation of vital information and the possibility of connection through all this hardship. At the height of social-distancing measures designed to curb the influenza pandemic, it was an enticing proposition that "a plain citizen . . . with his umbrella held aloft can walk along the street talking with . . . the house at the other end of the city," as MacFarlane described it.[11] Although the technology existed to modulate 650-foot radio waves so that they conveyed information, a broad understanding of how this feat actually worked still lingered far behind. Amateur experimenters like the MacFarlanes were successful in demonstrating the uses of wireless before its physics was fully understood. Research into electromagnetic radiation had recently begun to place radio waves along a spectrum that included visible light and X-rays. But it was the tatterdemalion ideas of the nineteenth century that still animated the popular mind. If MacFarlane performed like a magician refusing to reveal the secrets of his black box, it's because he didn't know how the trick worked.[12]

Through war and the pandemic, "the morbid effect of chasing voices at a great distance" was inseparable from visions of death and the afterlife.[13] Before the nature of radio waves had been explained, a prominent belief persisted well into the 1920s in the existence of an omnipresent, luminiferous ether: the invisible substance through which radio waves moved. (Surely waves must pass through some sort of medium in order to create ripples.) In the public reception of fantastic new communications technologies, scientific half-truths mingled with spiritual beliefs and occult practices. Often the activities of the two groups inspired one another, in what Friedrich Kittler has called a "relay race from illusionists to engineers."[14] For one observer at the time, wireless technology was "almost like dreamland and ghostland, not the ghostland of the heated imagination cultivated by the Psychical Society, but a real communication from a distance based on true physical laws."[15] Even the most technically minded inventors freely borrowed from the empirical and the paranormal alike. Oliver Lodge, an important figure in the identification of electromagnetic radiation and inventor of several key wireless technologies (the coherer and syntonic tuning), published a book entitled *Raymond, or Life and Death* in 1916, recounting his attempts to make contact with his son who had been killed in the war. In 1920, Thomas Edison announced that he was "building an apparatus to see if it is possible for personalities which have left this earth to communicate with us." Meanwhile, magazines instructed readers in curious techniques for increasing the reception of their own wireless devices, like holding hands and standing in an incantatory circle to form a human antenna.[16]

Today, it's not difficult to imagine that after enduring years of isolation and grief, people would feel the need for a device that could cast some pale reflection of face-to-face conversation. What a comfort it would have been to hear, for the first time, the rhythm of Morse code, or even voices skimmed from the air, and to know those messages meant that somebody—or some spirit—was out there. One of the magazines interviewing MacFarlane contained an editorial in the same issue that promised, "What a wonderful world it will be in one or two years hence! You will step out into the starlit night and myriads of voices—noiselessly and invisibly—will fill the air all over the continent, flung through the ether. . . . A perfect Babel of voices will greet you, but by means of your tuning devices you will be able to pick out the very voice you wish to listen to, tuning out all the others."[17] In order to join that "electronic kinship," all you had to do was take hold of a lamppost, lift an umbrella to the sky, and—like magic—hear the voices all around you.[18]

One hundred one years later, down the same streets the MacFarlanes sent their radio waves echoing, I was absolutely glued to my own pocket wireless device while trying to prepare for a pandemic whose shape seemed to evolve in real time. Standing with my partner in a grocery aisle in early March 2020, I unsuccessfully googled whether microwaveable or stovetop rice would be a safer bet if infrastructure maintainers could no longer report to work. Which utilities would last longer: electricity (delivered in Philadelphia by a private company) or gas (a municipal utility)? The next day, I tried to reassure myself that it was just paranoia that led me to download PDFs of medical supply checklists and CDC preparedness guides in case our internet connection also went down. "Who knew what was true," says the protagonist of *Severance* (2018), Ling Ma's uncannily timed novel of pandemic survival in New York City. "The sheer density of information and misinformation at the End, encapsulated in news articles and message-board theories and clickbait traps that had propagated hysterically through retweets and shares, had effectively rendered us more ignorant, more helpless, more innocent in our stupidity."[19]

The actuality of what communities experienced in the following months—checking on neighbors, delivering groceries, shuffling endless Zoom calls—was of course nothing like the dystopian imaginaries that led us to go with the stovetop rice over the microwaveable. (Didn't Snake Plissken cook over an open fire in *Escape from New York*?) Instead, the gradual accrual of new, socially distant habits became a hallmark of this pandemic's experience. Many of these habits were filtered exclusively through digital devices that conferred the privilege of distance for some—curbside pickup, contact tracing, touchless payment, delivery apps, remote work—while deepening the reliance on precarious gig work and continuous viral exposure for others.[20]

Yes, phones were able to lend a sense of ambient connectivity, a "perfect Babel of voices" to comfort us during the isolation of the Covid-19 pandemic. But after a century's worth of tech fixes to alienation and disconnection, to gadgets that have promised to inform, connect, and orient us in the world, those old dreams began to feel brittle. Digital devices were at the same time a pale approximation of in-person interaction, an uncommon privilege with over 50 percent of residents in some of my city's neighborhoods without internet access of any kind, and a fount of rampant misinformation about the virus and how to treat it.[21] In April 2020, thirty wireless towers were set on fire across Britain by people devoted to a conspiracy theory that 5G radio waves were making them more susceptible to viral infection.[22] People had finally had enough of the tech industry, but they turned to Luddism for the wrong reasons. Unlike the nineteenth-century British textile workers who destroyed the automated looms that had rendered their labor obsolete, these Luddites turned against the very machines generating their misplaced anger through fake news.[23]

During the influenza pandemic of 1919, pocket wireless seemed a fantastic dream, an achievement of the future. It was a solution imagined by many, adopted by few. But enterprising experimenters could at least reach out and touch its antecedents by strapping together some stovepipes, angling a couple of coat hangers just right, and lacing up their metal-heeled shoes. That imagined technology is undoubtedly now here, but it remains unevenly distributed and even more unknowable. I carry in my pocket, quite literally, over half the elements on the periodic table, extracted from almost every continent on the planet and compressed into a thin slab that allows me to dip my fingers into a stream of the thoughts and feelings of everyone I've ever met. I don't need to know how this gadget works in order to expect that it will. In countless facets of daily life, I wield one of the most complex and resource-intensive artifacts in history by relying on habit, know-how, tactility, and intimacy—varieties of what Susan Sontag calls "embodied intelligence."[24] These faculties are difficult to describe, though readers likely will already know what I mean. I mean the ghost stories you tell your friends about eerily timed ads and apps that must be recording your conversations. I mean the way you recognize a friend's texts by the rhythm of their notifications coming in, like the telegraph operators who could distinguish one another by the cadence, or fist, of their Morse code. I mean your tacit understanding that every time you search or swipe, text or like, the breadcrumbs left behind tell a story of your life, scattered somewhere across a combination of servers, a portrait in data whose completeness rivals the ways even you know yourself. The way we understand our friends, our selves, and the world around us is now determined by these unstandardized, emergent forms of tacit knowledge.

For most of us, the smartphone is like the thin black monolith in *2001: A Space Odyssey:* it supposedly makes us smarter just by touching it, but we have no idea how it works or where it came from.[25] During the 2007 unveiling of the iPhone, Steve Jobs said that "it works like magic," and in 2011 he introduced iCloud—connecting those devices—by explaining, "It just works." A peek behind the curtain would reveal a baroque chain of exploited assembly line workers, carbon-belching infrastructures, and toxic end-of-life scrapyards. To users, however, the magic of smartphones is that they are designed to feel weightless, immaterial, and transcendent—so much so that our bodies seem almost incidental to their futuristic existence. Cerium oxide slurries buff our screens flat to a level of industrial precision we are sure to scratch once we get our hands on them. As if disgusted by our fumbling touch, phones repel smudges with coatings described as oleophobic (fear of fat) and lipophobic (fear of oil) so that they bear no evidence of having been touched.[26] Fleshy human users must seem to phones like a necessary inconvenience, a presence they tolerate in exchange for the many things they seek to learn about us. As Shoshana Zuboff describes it, "We are exiles from our own behavior, denied access to or control over knowledge derived from its dispossession by others."[27]

The asymmetrical shape of technology today—how little we know of its material complexity, while it knows so much about us!—feels a certain way to users. After exposure to a century's worth of storytelling about what ubiquitous connectivity would be like, something seems off. A bargain was made with these surveillant, adversarial devices during an epoch of social distancing, as every happy hour and routine health care appointment, every first date and kindergarten class, moved online. Even though gadgets spewed misinformation and sowed political division while tracking our every movement, many of us relied on them to stay sane during the pandemic's isolation, shuffling in the process all previously settled distinctions between online and offline, distance and proximity, connection and touch. Before the pandemic, my phone was an inconvenience, the thief of my attention and imagination. It was something trivial that I could set aside and feel virtuous for having done so. But during the pandemic, I became tethered to my device in a qualitatively different way, acceding to its more ghostly and manipulative qualities in exchange for a semblance of social contact.

Is this how it was supposed to feel?

This is a book about the curious rituals of everyday users; the ways folk knowledge adopts complex devices; the uses of a tool that were never intended by its designers; and the role that fictions play in our tacit experience of technology. Every day, people negotiate a unique and intimate syntax with their tools, tacitly debating things like the relative merits of usefulness and efficiency versus habit and sheer aesthetic joy. Those micro-level techniques don't exist in a vacuum; in the aggregate, they add up to paradigm shifts in the history of technology through a series of wordless debates. Taken together, the everyday practices of users constitute an alternative philosophical tradition, a media theory from below, in which distinctly vernacular philosophies emerge that anticipate and sometimes exceed the conceptual frameworks sketched by a class of philosophers and writers who study them from afar. The chapters that follow take the high theory of technology as it was understood by thinkers like Marcel Mauss and braid it together with the low theory of users like us.[28] The late nineteenth century's age of machines up to the twenty-first century's age of tech inform this book, with each era furnishing its own lexicon of terms and concepts for capturing the embodied know-how of the end user. In recovering these minor and sometimes oppositional ideas on technology, my goal is to gather an actionable set of historical case studies portraying users as agents of technological change.[29]

Some of the case studies that follow form clear parallels with today's digital devices (like the pocket wireless); other comparisons will seem a bit further afield (like medieval plows and sailors' knots).[30] The range of sources animating this book is eclectic, an approach that reflects an efflorescence of new methods in academic disciplines devoted to the study of technology and culture during a period of rapid media change. Previously we relied on separate devices for separate forms of communication and expression, with clear sensory distinctions between each device-medium pair: photography and the camera, storytelling and the book, music and the stereo. Each medium in turn had a distinct academic discipline devoted to its study: art history, literary studies, and musicology.[31] Now that the smartphone has become our predominant means of access to media of all kinds, formerly rigid borders are colliding and collapsing, their differences becoming less distinct. (For example, it is a signature of our post-device condition that television seems no longer to refer to the furniture-sized box we plug in and turn on but rather a serialized narrative format. We use the same device to "watch TV" in fifty-minute episodes and "watch a movie" in a two-hour sitting.) Art, literature, and music's aesthetic forms are all flattened out into the figure of content, which is piped through phones that few of us actually use to take calls.[32] When all media belong to the smartphone—"the one device," as Brian Merchant calls it—each medium has less of a material or technical identity than it does a particular cluster of feelings or experiences associated with it.[33]

The user experience of the smartphone has shifted the terms in which media are studied. Over the past two decades, the convergence of countless forms of communication and cultural expression in one device has meant that academic disciplines previously devoted to medium-specific analyses are blending. Any interpretation of the past is irrevocably bound up with and enriched by present-day analogs. As these disciplines change in concert with their object of inquiry, researchers now borrow from a diversity of traditions in search of new models. It is now far less clear where the boundaries lie between "media devices, practices, objects, and texts embedded in our everyday lifeworlds," as Paul Frosh enumerates in his capacious poetics of digital media.[34] And so many scholars are increasingly turning their attention away from discrete devices and toward the *use* of media, with approaches ranging from philosophies of technique, habit, and know-how to histories of neglected ideas like psychotechnics, the useful arts, and the *chaîne opératoire*.[35]

What's interesting about this transition is that the experience of digital devices today surfaces a conceptual argument advanced by scholars for generations: that media have never had any fixed identities, and their meanings are as numerous as the people using them. Some researchers eschewing a focus on individual devices have returned to media ecology, an approach first pioneered in the 1960s suggesting that the ensemble of media used by individual people each day forms the perceptual and affective environment in which they live.[36] Already in 1988 Carolyn Marvin observes that late-nineteenth-century "media are not fixed natural objects; they have no natural edges. They are constructed complexes of habits, beliefs, and procedures embedded in elaborate cultural codes of communication."[37] Donna Haraway, writing in 1985's famous "Cyborg Manifesto," argues that technologies writ large can be partially understood as formalizations, i.e., as frozen moments, of the fluid social interactions constituting them. . . . The boundary is permeable between tool and myth, instrument and concept, historical systems of social relations and historical anatomies of possible bodies, including objects of knowledge. Indeed, myth and tool mutually constitute each other.[38]

If media, fluid and various, have always reflected their shifting cultural and historical milieus, then the interesting question becomes where users enter that continuum when they reach for a device. Every time users pick up a tool, its functional affordances (like storage, torque, or illumination) intersect with individual preferences, traditions, beliefs, and routines for using it.[39] The *chaîne opératoire*, or operational sequence, of any user interaction, as described by French archaeologist Andre Leroi-Gourhan, is made of these material and nonmaterial elements: "both gestures and tools [are] sequentially organized by means of a 'syntax' that imparts both fixity and flexibility to the series of operations involved."[40] In this sense, every user interaction is subject to an alchemical mix of technical details, cultural influences, and individual styles.[41]

The minute textures of the way we do things can in part be captured by analyzing the shape of the tools we use to do them. The pocket wireless was a technical ensemble: headphones, antenna, cables, microphone. We could study the physical details of its various iterations and what those devices might have been capable of. But pocket wireless was also a story people told themselves about disembodied presence and the future of ubiquitous connection. It was a cipher for politics, culture, and belief amid the twin traumas of World War I and the influenza pandemic. People threading wires through their clothes drew on a range of functional and fictional components when firing a telegraphic missive from aërial to ether and wondering who might reply.

By the time the MacFarlanes demonstrated their pocket wireless to the press, there had already been two decades' worth of storytelling, speculation, and experimentation to understand what ubiquitous connectivity would feel like. As early as 1899, when Guglielmo Marconi debuted his new wireless telegraph in the United States, the form factor of the pocket served as a kind of vanishing point for the ultimate development of the wireless transmitter and receiver. William Edward Ayrton described a feeling at the time that the world was "gradually coming within thinkable distance of the realization of a prophecy." If only that first clunky wireless telegraph could shrink down to pocket size, then two people could remain in constant contact no matter where on the planet they were. A friend could simply "call in a loud, electromagnetic voice, heard by him who had the electromagnetic ear, silent to him who had it not. 'Where are you?' he would say. A small reply would come, 'I am at the bottom of a coal mine, or crossing the Andes, or in the middle of the Pacific.' Or, perhaps, in spite of all the calling, no reply would come, and the person would then know that his friend was dead."[42] The future was portable, like in the very first advertisements for the iPod: "Every song you've ever owned. In your pocket."[43]

For twenty years, depictions of pocket wireless circulated in the form of baseless speculation, blueprints, prototypes showcased for the press, props and plot devices in fiction and film, and vaporware advertised alongside X-ray specs and vitamin tonics in the back of popular science and children's magazines. Over those two decades, countless words were floated to name and understand this imagined medium of mobile communication: aerophone, radiophone, Marconigram, aerogram, wireless telephone, radio. By the time the MacFarlanes revealed their own technique for speaking into the air, it had become a familiar sight to see someone walk down the street, reach into a pocket, and announce that a message had been received.[44] As Rebecca Tuhus-Dubrow writes in her history of the Sony Walkman, "The history of technology is in part the story of normal people starting to do things that used to be considered signs of insanity. First it was hearing the voices of people who weren't there; then it was tapping your foot to music nobody else could hear. And still later, it was talking when walking on the street, alone."[45] Techniques once considered mad, fantastic, or science fictional eventually become habitual and commonplace.

Rather than showcasing a procession of smaller and ever more powerful devices, the history of technology that underlies this book consists of countless users, each assembling a unique syntax from stories, moods, and historical circumstance. It reflects the tacit experience of those users and their understanding of how their magical devices worked by paying attention to the language they used to describe the way they did things. Much of this book is devoted to a philological premise: that words conceal tiny epistemologies, and that by tracing technical terms, jargon, and slang through the textual record, we can watch the evolution of a discourse on technology—and on the imaginative space between tools and their users. *A User's Guide to the Age of Tech* aims to resuscitate forgotten meanings buried within the words we use to describe how new technologies feel.

Up to this point, and for the sake of convenience, I've used words like *technology, device, tool, gadget, medium,* and *tech,* each of course easily recognizable. But even seemingly self-evident words evolve as quickly as the objects they name. In order to enrich the language we have at our disposal for questioning the directions new technologies take us in, I will ask you from this point forward to forget everything you know about words like *technology*. Today, we mostly use *technology* to refer to complex tools individually (a piece of technology) or collectively (military technology). In English, however, the word did not enter widespread use until the 1960s. Before then, it referred not to tools but to a specialized academic field devoted to the study of those tools. The term *technology* emerged "at a specific moment in American history—the 1840s," writes Clapperton Mavhunga, "when concepts like the useful arts and mechanical discovery, improvements, and invention became inadequate to describe steam power, electricity, the railroad, the telegraph, and myriad other new markers of 'progress.'"[46] During these industrial revolutions, technology was a precursor to engineering, and it transformed scientific principles into advanced machinery.[47] The "T" in MIT (Massachusetts Institute of Technology) for most of its existence would have been understood to refer to a specialized branch of learning on par with other -logies like biology, archaeology, and philology. It would take another century for *technology* to mean tools rather than their study.

Tech, meanwhile, is a word that today seems less and less to have anything to do with tools or the people using them. When economic observers say that "every business is now a tech business," they don't mean that more companies want to sell us gadgets. *Tech* instead refers to a business model that extracts value from our everyday habits, hopes, and routines in an effort to incorporate all human activities into the market. This approach was pioneered by the big five—Alphabet, Amazon, Apple, Meta, and Microsoft, companies so large that they sway the entire S&P 500[48]—but it has now inspired corporations of all kinds to capture and sell worker and consumer data: driver location for auto insurance providers, biometrics for athleticwear brands, shopper movements through the aisles of big box stores. All of this highly personal information is then sold to monopolistic third-party data brokers, which bundle information about us from countless sources as new commodities, as scholars like Mary F. E. Ebeling, Sarah Lamdan, and Armelle Skatulski have shown.[49] In the eyes of tech, the users of smart devices of all kinds—refrigerators, toothbrushes, watches, baby bassinets—are merely switching points, coincidental circuits meant to speed the flow of data, mined and sold by the barrel like oil. The word *tech* is an emerging shorthand for this process of capturing and monetizing data about the way we go about our daily lives.[50]

Both words are the product of vast economic revolutions and distinct markers of the moments they entered use. *Technology* fulfilled the nineteenth-century need for a concept that could describe the Industrial Revolution's labor struggles, as well as its annihilation of distance through steam and later electricity. Technology's dematerialized offspring, *tech,* names the twenty-first-century capture of patterns in everyday life and the extraction of those patterns as new forms of capital. I am concerned in this book with an intermediary concept. Between the broad upheavals of *technology* and *tech* is the far more subjective, imaginary, and creative realm of *technique:* the way we do things, not just the tools we use to do them. Throughout this book, *technique* is a concept that will give shape and meaning to transformations in the role that tools have played in everyday life over the past century. While the word *technique* does imply what most English-language readers will think it means, it also has a surprising range of other connotations that have fallen away over the same period of time.

Technique can of course be thought of as specialized procedures or methods taught in activities like running or painting: a runner shortening her stride length to decrease ground contact time and reduce the impact on her knees; a painter holding a brush like a stick to record broad arm gestures, or like a pencil to produce fine lines. *Technique* can also encompass unstandardized or tacit expertise gained through repetition and skill, like the old carpenter's trick of using a toothpick to get a screw's threads to catch in a stripped hole.

But in other languages, *technique* implies much more than procedure and skill. In addition to describing ways of doing things, its cognates in French *(la technique)*, German *(die Technik)*, and Italian *(la tecnica)*—to name a few—are words that refer to the much broader realm of "the human transformation of the material world."[51] Procedures and skills are joined by embodied know-how, stylistic differences between users, and the symbolic dimensions of tools. Complicating things further is the fact that each of these cognates in other languages can also refer to the tool itself that is used in a given procedure; the words can be translated into English using either *technique* or *technology*. Geoffrey Winthrop-Young, translator of Bernhard Siegert's influential work of German media theory, *Cultural Techniques*, notes at the outset that *Technik*'s "semantic amplitude ranges from gadgets, artifacts, and infrastructures all the ways to skills, routines, and procedures."[52] Eric Schatzberg notes further that academic disciplines like "'history of technology' in French is *l'histoire des techniques*, in German *Technikgeschichte*, in Dutch *techniekgeschiedenis*, in Italian *story della tecnica*, and in Polish *history techniki*."[53]

All of these words, from *tech* to *techniekgeschiedenis*, stem from the ancient Greek root *tékhnē*, a concept encompassing craft, art, skill, and making. "In its original, Aristotelian conception," writes anthropologist Tim Ingold, "*tékhnē* meant 'a general ability to make things intelligently,' an ability that depends upon the craftsman's or artisan's capacity to envision particular forms, and to bring his manual skills and perceptual acuity into the service of their implementation."[54] Scholars like Mavhunga have expanded the concept's contours even further, arguing that we should look beyond any settled, Western origins: "Nobody really asks: Where did the Greeks get that definition? Or: What did other civilizations, like the Egyptians for instance, call similarly denoted phenomena?"[55] Today, various styles of *technique* are prided in socioeconomic contexts requiring the maintenance and care of older infrastructures. Roopika Risam, for example, notes "a range of cultural practices that privilege making do with available materials to engage in creative problem-solving and innovation. These go by names like *jugaad* in India, *gambiarra* in Brazil, *rebusque* in Colombia, *jus kali* in Kenya, and *zizhu chuangxin* in China."[56]

In English, we haven't really had the benefit of a single word or concept that encompasses the way we do things, the tools we use to do them, and the question of why things have been done a certain way. Lost in the functionalist procedures of the English word is a more capacious understanding of *technique* as a form of culture, embodiment, politics, and even metaphysics.[57] In this expanded sense of *technique* for many languages, the how of procedure is joined by the why of culture.

The case studies in each of the chapters that follow seek to expand the range of what we mean by *technique,* opening up in the process a new understanding of digital devices that bridges procedures, know-how, stories, ideas, and tools. We will begin with the story of a 1941 conference in Nazi-occupied France devoted to the concepts of techniques and technologies—terms that were in flux amid the emergence of mass media and the rapid deployment of military industries of previously inconceivable destructiveness. The following chapter, chapter 3, leaves behind the philosophers in favor of what technicians know of their tools. The chapter is a study of the different kinds of tools referred to as *gadgets* and the changing reasons why. The evolution of the word *gadget* and its applications distills an entire discourse on tools and techniques by sailors, mechanics, and other technicians.

In comparing the high theory of philosophers with the low theory of technicians, we will see that technique represents more than simply ways of doing things. It also represents an array of ideas and arguments about the use of tools. These techniques become especially visible in moments of crisis, as we will see in chapter 2: the devastation of war shaped philosophers' capacity to understand technique past and present. Technique also comes to the fore during technocultural paradigm shifts, as it did for technicians navigating the transition from sail to steam ships or from crystal to transistor radios, as seen in chapter 3. In both cases, technique becomes visible when our habits are no longer our own.

Putting into conversation the ideas and experiences of philosophers and technicians provides a language for the know-how characterizing our user interactions today. Moving into the present, the book's final chapter, chapter 4, focuses on users. It asks: what does the user of a data-driven or computational platform know that is qualitatively different from users of tools that have come before? This final chapter returns to Philadelphia and explores new community-based cooperatives devoted to growing and maintaining their own internet infrastructures, showing that users are often more than just the passive recipients of new tools; instead, they actively shape the conditions in which those tools are used.

Describing technological change from the perspective of users is of the utmost importance during a moment when the tech industry continues to devise new methods for tracking our habits and hopes. In order to enrich the language at our disposal for describing the way new technologies feel and for critiquing the directions they take us, this book suggests a shift in how we value expertise: who gets to lay claim to it, and why it is important to diversify the forms of expertise that count in conversations on technology, privacy, and the public. Emphasizing technique in the age of tech, this book seeks to promote the agency of the individual life in a world that wants to reduce it to a collection of monetizable traits.

Alphonse Legros, *La charrue* (1874), etching. Yale University Art Gallery. Gift of G. Allen Smith.

2

The Way We Do Things

WHEN SOCIOLOGIST MARCEL MAUSS WAS INVITED TO present at a conference on techniques and technology in the summer of 1941, his initial instinct was to decline. "I'm not working on anything, not even for myself," Mauss wrote to a friend, Ignace Meyerson, a psychologist who organized the gathering in Toulouse.[1] For French Jews, this was a precarious time to consider travel, let alone the prospect of working. France had fallen to the Nazis in June 1940, and within months, the Vichy government, led by Philippe Pétain, began to pass anti-Semitic legislation. The October 3 Statut des Juifs systematically expelled Jews from any profession that could influence public opinion, including the military, media, law, and academia. Of four thousand Jewish university professors in France, 125 applied to the Ministry of Education for exemption from the Statut. Only ten were given permission to continue working, by Pétain himself.[2]

One of these ten was another invitee to the conference: medieval historian Marc Bloch, who received special consideration for his military service during World War I.[3] For Bloch, traveling to Toulouse meant the prospect of facing Lucien Febvre, his founding partner in the influential journal *Annales d'histoire économique et sociale*, who was also scheduled to present. Just months earlier, Febvre had forced Bloch to resign his editorship with *Annales* for fear that a journal with a Jewish codirector risked seizure by the authorities, emphasizing the fact that Mauss had agreed to anonymize all of his contributions to other journals.[4] Clearly while Bloch's right to employment was preserved, his positions were not. Bloch decided to make the trip south to Toulouse, although not primarily to attend the conference. It was just a stop-off before visiting the nearby University of Montpellier in search of a teaching job after the Nazi liquidation of his home university in Strasbourg.[5]

The position of Mauss, who at sixty-nine years old had decided to remain in the occupied zone of Paris, was even less clear. Within weeks of the Statut des Juifs, Mauss resigned from all of his academic positions.[6] In his resignation letter, he wrote, "In the current situation, it is pointless for me to be a target through whom the École [Pratique des Hautes Études] could easily be attacked. Though I fear nothing for myself, it is my duty to endanger only myself."[7] Amid food rationing, arrests, and deportations, many of Mauss's family, friends, and colleagues had fled the city. Mauss and his wife, Marthe, were evicted from their apartment when it was requisitioned for a German general. Though Mauss eventually agreed to send a paper to the conference to be delivered in absentia, he found writing it almost impossible. Suffering from diabetes made worse by the terrible conditions in the "cold, dark, and dirty" apartment they had been forced into, Mauss struggled with the physical act of writing: "I have no stenographer, otherwise I'd make an effort to dictate. . . . It's my thumbs that don't want to work."[8]

Meyerson and other friends continued to encourage Mauss, in part knowing the importance of his thoughts on the subject, but perhaps also simply to keep him moving. Meyerson implored, "Where is your study on the general theory of technologies? You know how it is anticipated, how it will be appreciated, how it will be read." Charles Fossey urged him, "As far as possible, finish your technology [essay] and get it published. And as soon as you're done, start something else. One must die with one's boots on."[9] Once it was completed and mailed off, the text of Mauss's presentation could very likely have been intercepted by surveillance officers carrying out the Vichy contrôle technique, which recorded the contents of about three hundred thousand letters and countless phone calls every week.[10] (Copies of Marc Bloch's correspondence were discovered by his biographer in the archives of the General Commissariat on Jewish Questions.[11])

During a time when the most basic aspects of traveling, writing, and surviving were made nearly impossible, a group of scholars committed to the study of history as an embodied process gathered to discuss technique. Meyerson, a psychologist interested in history, invited historians interested in psychology. Most of the conference presenters were associated in one way or another with *Annales*, the journal founded in 1929 by Bloch and Febvre. What became known as the Annales school of historiography was united by the desire to write history from below. Rather than emphasizing stories of great men, politics, and wars, the Annales school sought to examine the everyday lives of people going about mundane activities. They spoke of "mentalities," or a collective state of mind shared by a group of people during a given time period. They therefore drew on methods from the social sciences of geography, economics, sociology, psychology, and linguistics.[12] (Bloch, for instance, was greatly influenced by Mauss's uncle, Émile Durkheim, a famous sociologist whom he had studied with at the École Normale Supérieure.) Conference participants presented on the ethnography of labor, changing ideas of work throughout history, scientific practices, monastic rituals, and the routines of peasant life.[13] Diverse as their subjects were, each presentation was animated by the concept of technique.

This chapter traces the emergence of a theory, the lived experiences that were inseparable from the formulation of that theory, and its continued influence on contemporary scholarship. The 1941 conference in Toulouse was an important moment in the study of technology from the perspective of several humanities disciplines. Understanding the ideas put forward at this conference in their own terms will enrich the language we have at our disposal for describing how we use our gadgets today.

The French word *technique* is far more capacious than its English cognate. When Mauss and Bloch say *technique,* they might refer, as one would in English, to skills, methods, and procedures—that is, the way we do things. But they also use *technique* to refer to the tools that augment those actions. The bear claw we learn in the kitchen to guide a precise knife cut, fingertips curled under with knuckles against the edge of the blade: both gesture and knife here are *technique*. English translators render the term as *technique* or *technology,* depending on context. In translating Bloch's discussion of the two different shapes of farm plots commonly found in medieval France, for instance, Janet Sondheimer has the benefit of variety with English-language terminology. One Bloch's technique . . .

> the prime reason for the contrast between two types appears to lie in an opposition between two techniques [*l'antithèse de deux techniques*]

. . . is another Bloch's technology:

> the wheeled plough must undoubtedly be regarded as a creation of the agrarian technology [*une création de cette civilisation technique*] which ruled the northern plains.[14]

In 1930s French, this single word could slide almost imperceptibly between three different meanings: the way we do things, the tools we use to do them, and the broader technological systems formed by ensembles of those tools and their applications.

Bloch and Mauss both struggled to define *technique* in their writings before World War II. They stretched it beyond its common sense, at times trying to make the word do things it was never meant to do. For Bloch, technique was a real problem when trying to understand historical change. New techniques emerged throughout history due to a complicated mixture of political, cultural, and social factors. Given the limitations of historical methods at the time, however, the precise chain of causality between those elements was impossible for historians to reconstruct. No one could confidently say where innovations came from at different moments in time, or why certain tools and techniques were adopted in favor of others.

For both writers, the study of technique challenged easy distinctions between antiquity and modernity, prehistory and present. Detailed analyses of technical systems in the medieval world uncovered traces of ritual and magical thinking that could still be found in twentieth-century techniques. Archaic magic and modern technology should both be understood as a "series of facts" about doing things in the world, a young Mauss wrote in 1902: "Magic works in the same way as do our techniques, crafts, medicine, chemistry, industry, etc. Magic is essentially the art of doing things, and magicians have always taken advantage of their know-how, their dexterity, their manual skill" when using instruments like the divining compass, magic wand, and praying stick.[15] Bloch also explored the persistence of ritual and magic in *The Royal Touch* (1924), a book that asked how devout French Catholics could believe that a touch from the king's hand would cure disease. This and other scholarship by the Annales school begged the question: is it possible to clearly delineate the origin point of modern science and rationality? Or do our techniques show that traces of older social orders and belief systems still underlie even the most complex modern infrastructures?

While the term *technique* invited a range of ideas and approaches at the time, Mauss and Bloch agreed that it lacked the precision necessary to describe ongoing developments in modern machinery, weapons, and media. This desire for greater clarity in the study of modern humans and their advanced tools would be accelerated by the end of the war as new monstrosities emerged from its ashes. How could the same term encompass abacus and computer, throwing spear and atomic bomb?

And so technique was the focus of the Toulouse gathering in part because of linguistic necessity and in part because of historical circumstance. For years, the Annales circle had been discussing the need for a new approach, method, or discipline devoted to technique. Mauss wrote in 1927 that "we have never had the time and strength necessary to give technical phenomena the formidable place they deserve."[16] Bloch argued in 1932 that "nothing would matter more to the progress of our historical research than good work on the evolution of various techniques."[17] Febvre edited a 1935 issue of *Annales* devoted to the subject.[18] And Meyerson opened the Toulouse conference in 1941 by arguing that older fields devoted to the applied study of technique (physiology, scientific management, psychotechnics[19]) needed to be replaced by new historical approaches capable of exploring technique "from its first artisanal and rural forms to contemporary machinery."[20] The conference was the culmination of these long-running discussions.

Mauss and Bloch each had his own reasons for positioning *technique* as one of the most pressing concerns for the disciplines of sociology and history, respectively, but their approaches to the concept were not just the product of academic debates. Ideas are embodied things, especially when it comes to ideas about technique. Mauss first turned to the subject while running a cooperative bakery he founded with friends, La Boulangerie, in the belief that socialism expressed itself best through practice.[21] (Uncle Émile, who had urged Marcel to take his hands out of the dough and focus on his studies, was not pleased when the bakery went bankrupt in 1901.[22]) Serving in the army during World War I gave Mauss a very different perspective on technique when he noticed that English troops "did not know how to use French spades" or how to march to the rhythm of a French bugler. His was an ethnographic mind even during war; Mauss observed that "a manual knack can only be learnt slowly. Every technique properly so-called has its own form."[23]

Bloch left the trenches fascinated with the collective mentality of his regiment and the unreliability of his own memories of war, which seemed to him "a discontinuous series of images, vivid in themselves, but badly arranged, like a reel of motion picture film containing some large gaps and some reversals of certain scenes."[24] The daily procedures of warfare and the way they were so poorly remembered after the fact left Bloch with an understanding of history as a chaotic accumulation of individual psychologies, routines, and beliefs. Similar to memory, history was like "the last reel of a film which we must try to unroll," Bloch wrote in *French Rural History* (1931), "resigned to the gaps we shall certainly discover, resolved to pay due regard to its sensitivity as a register of change."[25]

Mauss also used cinematic metaphors around the same time to capture and describe the experience of technique in "Techniques of the Body" (1935), in which he noted that a particular style of walking, a gait, had become popularized by American films shown in France: "American walking fashions had begun to arrive over here, thanks to the cinema." In this famous essay, Mauss suggested that the body is the first tool we ever use: "I made, and went on making for several years, the fundamental mistake of thinking that there is technique only when there is an instrument."[26] But Mauss began wondering what it would mean instead to see the body as our "first and foremost natural instrument." Walking, swimming, yawning, pointing—each practice involves forms of imitation, creativity, repetition, and skill.[27] "The constant adaptation to a physical, mechanical or chemical aim (e.g., when we drink) is pursued in a series of assembled actions, and assembled for the individual not by himself alone but by all his education, by the whole society to which he belongs, in the place he occupies in it."[28]

Years later, a student of Mauss's, archaeologist André Leroi-Gourhan, criticized this idea as the product of someone without enough technical literacy to understand how tools work in the first place. In an interview, Leroi-Gourhan said that Mauss "discovered in techniques of the body a field particularly suited to the orientation of his mind: techniques without the burden of the 'trade,' the knowledge and the training of the maker."[29] Body techniques encouraged people to look at tools and see only themselves. But when we consider a tool as complex as the smartphone, the lack of technical detail in Mauss's concept of body techniques begins to seem like a feature rather than a bug. The fact that users (or sociologists, for that matter) do not possess a detailed understanding of a tool's composition does not mean they lack intimate knowledge of that tool through experience or observation. As Mauss wrote, "the tool is nothing if it is not handled."[30]

What would Mauss say of the way we scroll through social media feeds, losing ourselves in the flow of manual dexterity? To me, this is one of most interesting sites of technique today—so interesting, in fact, that I find it difficult to look away from other people's screens on crowded subways. Rest assured! My attention is never drawn by the content of the messages or the images themselves. (Anyway, it's impossible to make sense of what's displayed unless it's your own hands manipulating the speed and motion of the text.) Instead, it's the virtuoso gestures flicking above the glow that I find fascinating. I've seen people tap open an email and almost instantly swipe back to the inbox—leaving no time to actually read its contents—then mark the message unread and repeat: tap-swipe-unread, tap-swipe-unread, one by one down the inbox. Some move their lips while texting, silently mouthing the words they enter. Others rest the phone on their thigh and type with their index fingers, tapping so quickly that the screen seems hot to the touch; or they cup the phone in both hands and, between keystrokes, lift their thumbs so high that they look like ballistic missiles relying on autocorrect for guidance. Still others swipe through Instagram stories, skipping across accounts rather than tapping on the right to view each individual image of their friends' posts. What really gets me is the way some let their thumb linger. Rather than reading the image and then swiping to the next, some read while swiping. The thumb slowly drags, turning the image over gradually enough for the eyes to register while it moves. Like a solitaire player drawing a card from the deck, the hand slows itself just enough for the mind to match the card with its destination.

Similarly, I've caught myself scrolling and staring without actually seeing anything that passes by on the screen. It's as if I'm looking over my own shoulder, text flying by too quickly for me to understand anything beyond the gestures themselves. Each time I do this, the single-handed gesture is mechanically the same (edges of the device pressed between fingertips and fleshy part of the palm, thumb flicking upward) and technically identical (similar algorithms and servers queried). But it means entirely different things depending on the context. I could be killing time during the commute, catching up with a friend, or shutting my mind off amid a wave of information overload.

These incredible forms of dexterity are developed over countless hours of distraction, and they seem entirely unique to everyone. While on the surface we might assume that these gestures are habitual, unintentional, and involuntary, Mauss argued that all techniques are profoundly shaped by our cultural surroundings and individual dispositions alike. What is at stake for Mauss is not just an understanding of the ways we use tools; it is also the ways we use our bodies in the process. In body techniques, technical complexity is choreographed into the movement of the body and what it knows of those tools. "Man is an animal who thinks with his fingers," Mauss wrote later.[31] This is not to say that our individual techniques exist in a vacuum. A complex tangle of psychological, social, and material factors contributes to the way we do things. Every gesture has historical, social, and cultural dimensions (in a word, context) that members of the Annales school were hoping to reconstruct at different historical moments.

This sense of technique as a province of the body that is always already conditioned by material and cultural context has been echoed in countless canonical works of media theory since Mauss, like Jonathan Crary's *Techniques of the Observer*, a book about the "historical construction" of vision. Crary shows how nineteenth-century optical devices like the stereoscope and phenakistoscope were products of "a new arrangement of knowledge about the body and the constitutive relation of that knowledge to social power."[32] We can see this sense of technique more recently in Mara Mills and Jonathan Sterne's "taxonomy of techniques of listening." They write, "like any cultural practice, listening is composed of techniques, such as directive and diagnostic listening, sequential and mobile listening, auditory memory, auditory accommodation, and environmental sound interpretation."³³ Body techniques are interesting because they show how every aspect of our lives, down to the most mundane habits, are conditioned by and shared through social contexts. Even the most unique, lonely, seemingly socially distant technique is the product of these rich social connections.³⁴

~~~~~~

The 1941 gathering in Toulouse provided an occasion for participants to revisit their often intimately personal assumptions about war, memory, experience, and the use of new tools and techniques. Some participants directly addressed the precarious circumstances in which they spoke, while others used historical references to obliquely acknowledge the chaos surrounding them. For Marc Bloch, the conference was an opportunity to share one of the enduring themes of his scholarship: the idea that histories of individual habits and techniques can be assembled into a broad account of their cumulative effects. In other words, Bloch was a historian interested in asking how an individual user's techniques might be connected to a wider portrait of technosocial change throughout history. If Mauss devised a conceptual language for describing the micro-level techniques of the individual, Bloch described the mechanism of transmission between those individuals and the collective whole.

In Bloch's contribution to the conference, "Transformations of Techniques as a Problem of Collective Psychology," he asks why new techniques are met with resistance by some groups ("societies of routine") while they are immediately welcomed by others ("societies of invention").³⁵ He begins with examples of farmers encountering revolutionary change at two different moments in history: the agricultural revolution of the seventeenth century and subsistence farming during Germanic invasions in the sixth century—a significant point of reference given the context in which Bloch spoke these words.

In the first instance, the modern agricultural revolution was driven largely by crop rotation, which included planting clover and turnips during the winter months rather than letting the land lie fallow. These techniques spread throughout England in the seventeenth century and dramatically increased the amount of food the country as a whole was capable of producing. But France was slow to change its way of doing things. The individual French farmer didn't care about "increasing the productive forces of the nation." He was focused only on "keep[ing] his standard of living intact. . . . Most peasants, in a word, feared the great social boulder that seemed the inevitable consequence of new methods." These new techniques were met with the resistance of tradition on intergenerational farms where "obstinate routine" had become deeply engrained.[36]

But over a millennium earlier, farmers of the sixth-century Merovingian period seemed to have no problem quickly adopting rye as their new staple cereal grain when it was introduced through foreign invasions. "Substituting rye for wheat or barley did not affect the social system," and during a time of invasions, raids, and forced migrations, these farmers had little to lose. Bloch, once again gesturing to the setting of the conference in Vichy France, adds: "One has the feeling (I dare not say more, for the moment) that the terribly tragic conditions of social life then were favorable to innovations." Taking a long, comparative historical view, it seems that periods of relative peace and stability, as in the seventeenth century, allowed for inflexible traditions to take root.[37]

For some time Bloch had been arguing that economic, scientific, or mechanical answers to questions of technical change were not in themselves sufficient. Craftsmen, artisans, scientists, and engineers all invent in different ways, and societies accept or slow the adoption of these new techniques for complex reasons. To understand why, historians must attend to "the internal structure of society, the interactions of the diverse groups which form it." In his conference paper, "Transformations of Techniques," Bloch critiques economist François Simiand for tying the appearance of new inventions to top-down economic incentives and scientific developments. And in a different context, Bloch levels a charge that today we might call technological determinism at historian Richard Lefebvre des Noëttes, who advanced the pernicious argument that the invention of the harness and saddle was responsible for the end of slavery because it reduced the need for forced human labor in the fields.[38] Bloch departs from these monocausal explanations and instead situates technical facts within their sociocultural milieus.

In his earlier influential work *French Rural History* (1931), Bloch considered an argument common in the field at the time—that differences in the shape of farm plots are related to differences in types of plow—before slowly deconstructing the whole idea. The *araire* was a type of plow that, because it lacked wheels, was easy to turn and pivot as it dragged across the soil and so produced irregular and open field shapes. A different type of plow mounted on wheels, *la charrue*, in contrast, was difficult to turn, so it was associated with long, straight fields. Bloch was unsatisfied with cold, technical explanations like these:

> Are these purely material factors *[considérations matérielles]* a sufficient explanation? It is certainly very tempting to trace the whole chain of causation back to a single technological innovation *[invention technique]*. . . . But we must be careful; such reasoning would fail to take account of the thousand and one subtleties of human behavior. It is true that plots had to be long if they were to accommodate the wheeled plough; but did they also have to be narrow? There was nothing in practice to prevent the occupants from dividing the land into a smaller number of plots, of sizeable length and breadth, so that each holding was composed of a few well-shaped plots instead of a host of very narrow strips.[39]

The medieval plow demonstrates for Bloch how the choice of a technique is always a social choice, one that exists in dialogue with the material details of the tools used. A later generation of early twenty-first-century scholars, loosely grouped under the heading of German media theory, would echo this idea in describing the plow as the foundational, symbolic act of politics. For Bernhard Siegert, "as ancient sources attest, plows were used to draw a sacred furrow to demarcate the limits of a new city . . . [marking] an inside domain in which the law prevails and one outside in which it does not," and for Cornelia Vismann, "the agricultural tool determines the political act."[40] In their writings on cultural technique *(Kulturtechniken)*, these theorists highlight an academic bias toward understanding culture primarily through text, rather than objects and techniques in all their material and technical detail. "Isn't it odd," write Sybille Krämer and Horst Bredekamp, "that the historical semantics of 'culture' refers back to agrarian methods and operations and to hand-based crafts? 'Culture' has its largely prosaic origins in the tilling of a field *(cultura agri)* and in gardening work *(cultura horti)*; it is first and foremost the work with things—their cultivation—that surround us on a daily basis."[41]

Bloch was an antecedent to these ideas in German media theory and to their understanding of culture as proceeding (etymologically and symbolically) from agriculture—as well as to the countless negotiations it involves between tool, technique, and society—rather than from text alone. Bloch was also a thinker who believed that in order to understand and appreciate these choices of technique, one had to be equally well versed in social structure and technical detail. In the words of Febvre, Bloch was determined "not to be the kind of historian who writes about agriculture in ignorance of what is an ox, a wheeled plough or a crop rotation."[42] What Bloch described in his presentation as "the importance of technical facts *[l'importance des faits techniques]*" for understanding historical change took on special meaning during this wartime conference, amid the machineries of war, in a society that he felt was "haunted by technique."

Mauss deemphasized tools in favor of a micro-level attention to the bodies using them in "Techniques of the Body." By 1941, however, his interests had begun to scale up to the macro. His Toulouse paper, like Bloch's, was preoccupied with the question of how the techniques of individuals add up to broader affects on the scale of civilizations. His conference paper, "Techniques and

Technology" *(Les Techniques et la Technologie)*, cleared the ground for future scholarship on these questions. While other conference presenters formulated their approach to technique in negative terms (i.e., how it should *not* be studied), Mauss was programmatic in his essay. He suggests an outline for a new collaborative academic discipline—*technologie*—using a term that was a relatively recent invention. (None of the other participants use this word in their presentations.) And he begins with a definitive statement on technique. "At a time when techniques and technicians are fashionable—in contrast to so-called pure science and philosophy, accused of being dialectic and sterile—it would be necessary, before extolling the technical mind, to know what it is."[43]

Like Bloch, Mauss had scattered the subject of technique throughout his scholarship up to that point. He created a recurring section on technique in the *Année Sociologique*, the foundational journal established by Durkheim. He was also one of the few professors in France to have offered a university course on the subject.[44] All of this meant that the participants of the Toulouse gathering were especially eager for a statement from Mauss that synthesized his thinking on technique. The paper that resulted was deeply personal. Mauss evokes friends who had recently passed: Henri Hubert in 1927 and François Simiand in 1935.[45] He alludes to the wartime food rationing that Parisians were being subjected to.[46] He reaches back to memories of former teachers, and to the books his uncle gave him when he was young, including popular works like *Les Merveilles de l'Industrie* (The Marvels of Industry, 1870) by Louis Figuier and Edmond Becquerel, both scientists who became popularizers of scientific and technical knowledge.[47] Woven throughout Mauss's nostalgia for a time that seemed each day to recede farther into the past were speculative notes on the future of technique both as a thing in the world and as an object of study.

The Toulouse essay begins with a definition of technique that distinguishes it from what we practice in the arts or in religion. (Throughout his work, Mauss uses the word *religion* interchangeably with *ritual* and *magic*.[48]) For Mauss, technique is an ensemble of movements or actions, in general and for the most part manual, which are organised and traditional, and which work together towards the achievement of a goal known to be physical or chemical or organic. This definition aims to exclude from consideration those religious or artistic techniques, whose actions are also often traditional and even technical, but whose aim is always different from a purely material one and whose means, even when they overlap with a technique, always differ from it.[49]

The goal of a technique—peeling an apple, sinking a screw—is primarily mechanical in nature. In other words, there is a direct relationship between means and ends. Turning an apple beneath a small paring knife and the thumb is much more efficient than trying to peel it with a big chef's knife on a cutting board. But an artistic act is guided by creativity, beauty, and style—in a word, aesthetics. The weight of a line in a drawing is guided by the individual experience and values of the artist, which feed back on the process and product of that artwork. As John Taggart remarks, "Technique is the way art means, and sometimes it thinks back, meaning more than we meant it to."[50]

Magic also resembles technique, if only superficially. Both magic and technique pair an instrument with particular kinds of gestures. Both demand craft and dexterity, ranging from magic tricks using the deceptive practices of card manipulation to the black kyanite crystal that cleanses a room of bad energy after difficult social interactions. The core difference with technique for Mauss is that in religious, magical, or ritual acts, "it is not really believed that the gestures themselves bring about the result."[51] In technique, there is a direct relationship between cause and effect—"food is cooked by means of fire"—as opposed to magic's ritual correlations—"a man stirs the water of a spring in order to bring rain."[52] I can't speak for crystals, but at least in the case of cards, people know it's a trick.

This is a far different dividing line between technique and magic than the one famously drawn by Arthur C. Clarke with his Third Law: "Any sufficiently advanced technology is indistinguishable from magic."[53] For Clarke, techniques seem magical when we can't understand how they work. The kind of magic described by Mauss, however, contains a shrewd mixture of belief and understanding. Magic for Mauss "is a kind of language. It translates ideas."[54] What matters is not whether we can explain, read, or interpret the techniques; instead, what matters are the effects of our belief in the process of performing them. Magical gestures are still effective; they still do things, if we understand them as creating forms of social cohesion and shared experience.

By Clarke's lights, most of the devices and infrastructures we rely on today would be described as magical, yet we don't stand in awe of this magic. Instead, we develop rituals to manipulate it ourselves. Anthropologist Nicolas Nova has cataloged some of these curious rituals: holding an arm overhead while on a call to get a stronger signal, tilting a phone to calibrate its GPS.[55] (This is not too different from the early radio listeners of a century ago, who stood in a circle around their sets, holding hands to get better reception!) These rituals extend from hardware to software as well. Think of the way that invisible social media algorithms can manifest themselves as everyday feelings. We think: why didn't my post get any likes? Why am I being shown this ad now? Media theorist Taina Bucher has written of an "algorithmic imaginary," or the mental models, folk theories, and forms of "practical knowing" that users devise to operate the hidden processes determining their social media feeds. One specific technique she describes is "conscious clicking," in which users try to influence their newsfeed content by clicking only on the things they want to appear. Users who consciously click "disrupt their 'liking' practices, comment more frequently on some of their friends' posts to support their visibility, only post on weekday nights, or emphasize positively charged words."[56] Whether or not conscious clicking actually does what users hope it will, the practice still ends up affecting the broader algorithms that learn from user behavior. Right or wrong, we understand ourselves through these rituals, and they feed back on our understanding of the world around us. Functionality, belief, and social cohesion are inextricably linked.

A technique is never solely functional, solely aesthetic, or solely ritual. While the distinctions are conceptually useful for Mauss, they are nearly impossible to identify cleanly in practice; nor would we want to. Even the nineteenth-century sources Mauss drew on combined these categories in their writing. While the fine arts were often contrasted with what was known in the nineteenth century as the useful arts, these categories were already shot through with contradictions. The activities understood to be part of the useful arts—writing, printmaking, plumbing, agriculture, locomotion, medicine—spanned a contentious mixture of manual skill and mechanical automation during a period of rapid industrialization. Several writers tried to import the language of the arts into these mechanical discourses in order to elevate them. Andrew Ure was a Scottish scientist who, in *The Philosophy of Manufacturers* (1835), imported discourses from the fine arts into the useful arts in order to elevate industrialists like factory owners, inventors, and engineers. He granted them "a degree of creativity denied to artisans, likening the cotton mill to 'individual masterpieces' that are studied in the 'philosophy of the fine arts.'"[57]

Technique, art, and magic also commingle today, but in different proportions. Mauss's essay is a compendium of nineteenth-century thinkers, each of whom we could imagine categorizing our digital techniques in quite different ways. Outlining a theory of technique began for Mauss with a number of idiosyncratic works by English, French, and German writers that never coalesced into a full-fledged academic discipline. These diverse lines of influence included English archaeologist Augustus Pitt-Rivers, whose museum Mauss had visited years earlier, referring to him as an "encyclopaedist."[58] Pitt-Rivers was most famous for his typological displays of stone tools, shields, throwing spears, and other artifacts. Because he was working in the 1880s, before methods like radiocarbon dating were possible, Pitt-Rivers arranged objects along an imagined historical timeline according to their form, assuming that objects with greater ornamentation must have been created more recently.[59] Mauss also cited German engineer Franz Reuleaux, who opened the first technical schools in Germany and developed a philosophy in *Kinematics of Machinery* (1875) that sprang from his hands-on experience as an engineer.[60]

An important yet indirect influence was Ernst Kapp, a socialist exiled from Germany after participating in the European revolutions of 1848 who settled in Texas during the years leading up to the Civil War. Kapp founded an abolitionist group in 1852 called the League of Free Men (Bund freier Männer) and worked in an agricultural commune among other German exiles, writing there over a period of twenty years his *Elements of a Philosophy of Technology*, a work that described tools as a form of "organ projection." Kapp drew on Aristotle's description of the hand as the "tool of tools," a play on the interchangeable use of the Greek word *organon* to mean both tool and organ.[61] One of Mauss's teachers, Alfred Espinas, borrowed heavily from Kapp in his book *Les Origines de la Technologie* (The origins of technology, 1897).[62] Like Durkheim, Espinas was a figure at the beginnings of sociology in France, and the two had a contentious, competitive relationship. Durkheim even advised his young nephew to keep a distance from Espinas.[63] Drawing primarily on literary sources from ancient Greece, Espinas suggested in *Les Origines de la Technologie* a philosophy of action he called praxiology. For Espinas, *technologie* was the way users applied and organized a sequence of techniques, while *Technologie*, with a capital "T," named the generalized principles that a scholar could deduce from these actions. Guiding these considerations was praxiology itself—above all, "the most universal forms and the highest principles of action among a group of living beings."[64]

In his conference paper, Mauss synthesizes these nineteenth-century influences into a new program for the study of technique once the war is over. He uses the term *technologie* to describe this speculative discipline. Before *technology* referred to things like machines or electronics, it signified a way of studying and thinking about tools. In the same way that psychology is the study of the psyche and musicology is the study of music, technology for

Mauss was to be a new field of study devoted to the history, culture, and psychology of techniques. If technique is the way we do things, then technology asks what it means that things are done a certain way.

Technology—this proposed "science dealing with techniques"—was to include psychologists, historians, sociologists, and ethnographers. But Mauss also hoped that the field could involve "the recruitment of workers and technicians" who could lend their craft knowledge to the discipline. Technology could be a field where scholars and technicians collaborated on the classification of different techniques across industries, arts, and crafts. (We might today imagine a comparative analysis of body techniques in sports, motorsports, and e-sports, for example.) In working toward a "descriptive technology," Mauss lists the different types of techniques that scholars could draw on:

1. Sources that are classified historically and geographically, such as tools, instruments, machines: the last two being analysed and assembled.

2. Sources that are studied from physiological and psychological points of view, including the ways in which they are used, photographs, analyses, etc.

3. Sources that are organised according to the system of industry of each society studied, such as food, hunting, fishing, cooking, preserving, or clothing, or transport, and including general and specific usages, etc.[65]

Each of these techniques would be studied as "characteristic of their social condition." In other words, the way users unlock their phones would be seen as a dialogue between individual users' proclivities, the raw materials they use to perform that action, and the social norms in which they are embedded. Echoes can be heard here of Bloch's understanding of technique as a negotiation between technical innovations and social tradition.

Applying this approach to the world in which he wrote, Mauss observes a complex assemblage of techniques now inextricably linked. The challenge for a *technologie* of 1941 would be to disentangle these "machines used to produce in series even more precise and immense or compact machines, which themselves serve to manufacture others, in a never-ending chain in which each link is built. . . . Even the most elementary technique, such as that relating to food production . . . is becoming integrated in these great cogwheels of industrial plans." His speculative notes toward "planning" at this essay's close suggest a *technologie* in which scholars and policymakers might better coordinate the ensemble of machines, individuals, and societies that were now bound to one another:

> The coordination of the ensemble of these projects cannot be left to chance. Techniques intermingle, with the economic base, the workforce, those parts of nature which societies have appropriated, the rights of each and or everyone—all crosscut each other. . . . More than ever, to speak of a plan is to speak . . . of morality, truth, efficiency, utility, the good. . . . It is pointless to contrast mind and matter, ideal and industry. In our times, the power of the instrument is the power of the mind and its use implies ethics and intelligence.[66]

Once the war ended, Mauss hoped that *technologie* could become "the activity of a people, a nation, a civilization" that plans for the most ethical coordination of infrastructures and resources possible.[67] It is a program for action that, in the absence of any functioning regulatory policy on tech, we are still waiting for today.

The conference in Toulouse featured what would be the final works of several of its participants. After months of communication, Bloch was able to secure a teaching position at the New School in New York. Tragically, only four of his six children were granted visas by the American consulate in Lyon. He had no choice but to remain in France. After presenting in Toulouse, "Bloch outlined projects he had scant hope of publishing,"[68] and a year of uncertainty followed for him and his loved ones. Sometime in 1943, Bloch joined the French resistance. He became a regional organizer, edited a major underground newsletter, *Cahiers Politiques*, and used archival research as an excuse for permission to move about the country.[69] In March 1944, he was arrested by the Gestapo in a small apartment above a dressmaker's shop in Lyon, where he was found working amid stacks of papers and a radio set. After being tortured for months, Bloch was finally executed by a Nazi firing squad along with twenty-six others on June 16, 1944. He was fifty-six.

Paris was liberated in August. By November 1944, Mauss was able to return to his apartment and to his position at the Collège de France.[70] But he rarely communicated with colleagues after that, receiving visits only from his nieces and nephews, friends, and former students. According to Claude Lévi-Strauss and other former students who visited, Mauss deteriorated rapidly and was unable to recognize his old friends. Around the time that Meyerson published Mauss's "Techniques and Technology" essay among the proceedings of the Toulouse conference, his brother moved in to care for him. Mauss passed away in 1950 at age seventy-seven.

It wasn't long before the ideas shared at this groundbreaking conference were taken up by other thinkers. Technique was a concept that shed a new light on existing topics of research like agriculture and ritual, challenging scholars to attribute agency and causality to people, objects, and processes that would previously have been overlooked. At the same time, technique itself became a new object of study, a constantly moving target that set the stage for new kinds of arguments about belief and social change. These ideas cleared the way for a new generation of scholars in France who developed what are now accepted as some of the classic frameworks of media theory and the history of technology. These ideas flourished in the 1950s and 1960s among the likes of André Leroi-Gourhan (who described the ways that our prehistoric human ancestors evolved alongside their tools), Bertrand Gille (who argued in a two-volume *History of Techniques* from antiquity to the present that the development of technical systems outpaces and determines the development of cultures), Jacques Ellul (who argued that the rationality and efficiency of technique had eclipsed all other forms of social belonging in modern life), and Gilbert Simondon (who elaborated a theory of individuation: the way individual subjects are continuously produced, differentiated, and combined through technical assemblages).[71] Many of these ideas are only recently gaining influence among Anglophone scholars as more work begins to appear in English translation.

At the time, the promise of a comparative *technologie* that analyzed techniques from antiquity to the present led Mauss to wonder how the history of science might be imagined differently if it began from the perspective of technique. Just how different are the pure reason of scientists and the practical reason of workers and technicians? What if studies of mathematics began with the techniques of counting? This latter question is now echoed in discussions of elementary cultural techniques by German media theorists, who frequently reference the basic human practices (like humming a tune or sketching a figure) that existed long before broader cultural forms (like music or fine art).[72] Mauss writes, "The oldest calendars are as much the work of farmers as of religious minds or of astrologers: technique, science and myth are there blended. In the same way, pigeons had been selectively bred before Darwin found the notion of natural selection."[73] The brilliance of technique can be found in the inexpert gestures of the most distracted user just as readily as it can in the philosopher describing those gestures from afar.

Today, technology is a thing studied rather than a field of study. But even if the terminology that animated scholarly debates on technique circa 1941 seems strange, the underlying ideas still feel fresh in a world that continues searching for ways to make sense of everyday technological upheavals. Bloch and Mauss were comparativists who drew on methods from geography, archaeology, history, psychology, economics, and sociology, asking how scholars in the humanities and social sciences could approach cold, mechanical facts of instrumentality and efficiency in ways that technicians couldn't. By 1941, the nineteenth-century sources they drew on had evolved into applied technical fields like engineering. The original projects of philosopher-technicians like Kapp and Reuleaux included cultural, historical, and ethical dimensions that were gradually abandoned during the rise of technical universities and departments of applied sciences. Mauss in particular rescued those elements from the earlier sources and used them to advocate for a humanistic technology criticism. This move is especially relevant at a time when computer science and cultural theory are coming into ever more frequent conversation through natural language processing, computer vision, and generative AI. As people begin using computers to write for them, we will need to decide how to reintroduce forms of historical, theoretical, and technical knowledge to one another—forms that have been kept separate for generations.

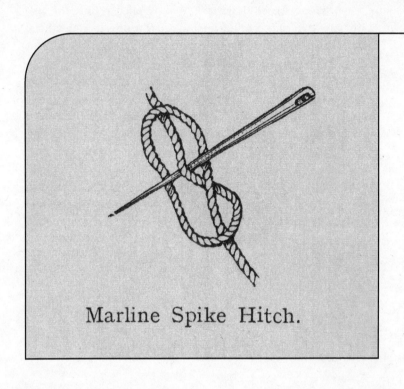

Illustration of a marlinspike hitch. From *Manual of Seamanship* (London: Sir Joseph Causton & Sons, 1908). Photographed by Digital Imaging Studio, Princeton University Libraries.

3

Technically Speaking

FOUR DECADES EARLIER, IN 1899, A DIFFERENT KIND OF debate on the way we describe tools was taking place in the pages of *Notes and Queries*. A British journal of amateur scholarship on history, literature, and folklore, *Notes and Queries* announced itself on the masthead to be "a medium of intercommunication for literary men, artists, antiquaries, genealogists, etc." In the June issue, a correspondent who gave only their initials—W.F.R.—posted a query on this Victorian bulletin board:

> A trial recently took place at Bristol . . . for damage done to a "gadget." The word does not occur in Halliwell, Smyth's *Sailor's Word-Book*, or the *H.E.D.* It evidently is the name of some kind of boat, which in the present case was used for the discharge of vessels in the harbour. Can any correspondent kindly give an exact definition, and also suggest its history and probable derivation? Is it local, or an importation, or a new coinage?[1]

Unfortunately, W.F.R.'s question went unanswered. The thread wasn't picked up again for almost twenty years, when an Archibald Sparke posted another query on *gadget* in this Urban Dictionary of the British Empire:

> I expected to find this word in the *Sailor's Word-Book*, as I had an indistinct recollection of hearing it at sea, but I cannot find it in Smyth's compendium of nautical terms; neither is it in the *N.E.D.*

or the *E.D.D.*, or Farmer and Henley's *Slang and its Analogues*. Can any one say in what connexion it is used?[2]

This time the replies began to roll in, complicating W.F.R.'s understanding of *gadget* as "some kind of boat." One reader reported that *gadget* "is a colloquialism in the Navy for *any* small fitment or uncommon article—for example, 'a curious gadget.' I never came across anybody who could give a derivation." Sparke followed up on his original post to say that someone had since sent him "a list of words and phrases used by our soldiers at the Front." (At the time of this correspondence—October 1918—the Great War was in its final months.) For the soldiers, "its meaning is given as billets or quarters of any description, and sometimes it is used to denote a thing of which the name is not known."[3]

The issue sparked conversations at a meeting of the Devonshire Association, a learned society and social group frequented by the kind of people who would read a journal like *Notes and Queries*. Members debated whether *gadget* should be recorded in their "list of local verbal provincialisms." But a naval officer present at the meeting explained that it couldn't have originated in any one town in England. *Gadget* was a word that belonged to the sea, carried around the globe by sailors of countless languages and cultures. For these sailors, *gadget* was used to name "a tool or implement, the exact name of which is unknown or has for the moment been forgotten." The conversation continued over several issues, but in the end, no definitive answers were agreed on. As more examples of gadgets came in, the exchange proved circular. What does the word *gadget* mean? Anything you don't know the word for. What kind of thing? It depends. One person hears a rumor that gadgets are little boats, another that they are private quarters. Some claim that gadgets tend to be implements and tools, others articles and objects.

While "literary men" chattered across what they described as their "telegraphic lines of literary communication" trying to pin down this specimen and discern its origins, sailors and other technicians were freely using the word *gadget* in both practice and print.[4] The "Queries and Answers" column of the American magazine *Marine Engineering*, for example, contained a July 1900 letter from a reader who sought the best way to connect the different valves of their boat's engine. The editors responded: "If your engine has a linking in gadget on the I.P. [intermediate pressure] valve gear, it would be well to link it in a little."[5] Amateur folklorists hunted for what they thought might be the unique origins of the term in small English villages. But *gadget* had long ago spread across the Atlantic. For which side of the ocean the word was an import or an export is hard to say, but it's clear that by the time the antiquarians of *Notes and Queries* realized that a gadget was a thing, the slang had already moved on. If you have to google it—*bet, shook, drip*—it's probably already over.

Gadget's migrations across the globe brought semantic changes along the way. With the rise of commercial steamships and private motorboats replacing sail, *gadget* passed from sailors to mechanics and engineers, and from there to pleasure boaters, automobile owners, and airplane pilots in the opening decades of the twentieth century. New communities of practice adapted the word to fit different needs, and its connotations changed along with the different tools and techniques to which it was applied. A word that was originally fashioned to refer to simple ropes, pulleys, hand tools, and other implements aboard sailing ships had begun to signify individual components within the complex machinery of steamships and motorboats, as it did for the editors of *Marine Engineering*. Within a couple of decades, it was more common to hear *gadget* refer to mechanical cogs and gears than to simple hand tools, as in "He was an engineer, in fact,—a man who knew every nut, bolt, and gadget of his submarine" (1918), or "unscrewing nuts, looking at carburetors, examining spark plugs, and testing aim-pump valves or any other gadget" (1919). The slapstick characters in Rudyard Kipling's story collection *Traffics and Discoveries* (1904) speak of the "steamin' gadgets" on their ships.[6]

So much had changed in the interval between these two queries on the meaning of a word, from the late nineteenth century to the end of World War I, from the last days of sailing ships to the beginning of aerial combat. "A generation that had gone to school on horse-drawn streetcars now stood under the open sky in a landscape where nothing remained unchanged but the clouds and, beneath those clouds, in a force field of destructive torrents and explosions, the tiny, fragile human body," noted Walter Benjamin.[7] Meanwhile, new cultural forms emerged to explore the affordances of recorded sound, moving pictures, and voices present at a distance. In the process of these sweeping social and technological changes, words evolved alongside the tools and techniques to which they referred.

In this chapter, I tell a story about one of these words and its circulation in print networks and wireless messages, in slang and sailors' yarns, in techniques and their description. *Gadget* was a term of art that formed an essential part of the nineteenth-century sailor's tool kit. The word itself was a kind of tool that only meant something when applied in a particular context—a privileged example of the philosopher Ludwig Wittgenstein's dictum "die Bedeutung eines Wortes ist sein Gebrauch in der Sprache [the meaning of a word is its use in the language]."[8] Since Wittgenstein, countless disciplines devoted to the history and philosophy of language—pragmatics, lexicography, historical ontology, historical linguistics—have explored the metaphor that words are like tools. We pick words up and use them to do things. But words referring to actual tools and techniques pose sticky questions for these language disciplines.

When we look into the history of the word *gadget*, we're reconstructing not only the different connotations of that term through time but also the specialized tools, techniques, values, and know-how of the communities who put the word to different ends. These practices are far more difficult to reconstruct than the words used to describe them. Amateur antiquarians and even professional lexicographers interested in nineteenth-century seafaring were outsiders looking in circa 1918, reassembling into stories what faint echoes they were able to collect of these rich practices. The echoes form a kind of dialogue between the shipboard technicians and the deskbound antiquarians who tried to capture a portrait of their practices, but it was a dialogue of a very limited kind; as Toril Moi notes, "The less we know about a certain region of practice (cooking, sewing, car mechanics, bullfighting), the less we will get out of the dictionary's presentation of its terms."[9] When we consider the language of sailing, what's the use of a landlubber's knowing that *marline* or *marling* (the spelling varies) refers to thin twine woven from two strands of hemp and coated with tar? We could further learn that its referent is a type of *small stuff,* or that it is used for *seizing* (binding two ropes together) and *serving* (wrapping the ends of a rope). But would we understand why ropes fray, slip, disintegrate, and how seizing or serving with marline can preserve them?

Gadget, as a word referring to tools of all kinds, provides insight into the ideas and values of the people who used those tools, and because it has been applied to objects ranging from nineteenth-century ropes and pulleys to twenty-first-century smart devices, the word distills an evolving discourse on tools and techniques. It offers a point of entry into the many regions of practice where it was applied. As we will see by using *gadget* as a case study, the semantic drift of the language that people use to describe their tools leaves behind a record of their ideas about technique. If *technologie* for Mauss referred to the study of techniques, then users, craftspeople, technicians, and others with practical knowledge were just as much technologists as the philosophers were. But while the philosophers in chapter 2 studied tools and techniques as the notional "relics of bygone instruments of labour," the technicians we will meet here actively generated concepts about tools in the process of using them.[10] The word *gadget*—as a placeholder that sits somewhere between a particular tool and a user's techniques for using it—evokes those historical techniques while also highlighting our inability ever to fully recover them.

Spoken among sailors for decades, the word *gadget* first began to appear in print around the 1880s. By this time, however, the tools that the word traditionally described were already antiquated. With new steamship technologies powering the consolidation of colonial empires around the globe, the oceans of the 1880s were crisscrossed by heavily traveled shipping lanes and increasingly affordable transatlantic passenger transport. An ever-growing web of underwater telegraph cables was reinforced by an extensive, shipboard postal network and packet trade. The nautical tools that now captured the public's imagination were dredgers that returned samples of seabed sediment and strange new creatures, or sounders that plotted prospective courses for transatlantic cables, in the process creating the first maps of the ocean floor's topography.[11] By 1899, the year that W.F.R. first inquired into *gadget*, Guglielmo Marconi was demonstrating his ship-to-shore wireless telegraph. (The first wireless message ever sent in the United States was reportedly, "R U there?"[12]) In so many ways, the world's vast oceans were becoming increasingly traversable.

But as sailing ships struggled to compete with the increasing dominance of steam-powered vessels in transatlantic commerce, nautical fiction featuring sail was as popular as ever. According to maritime historian Helen Rozwadowski, "As late as the 1880s and 1890s, when steam had largely supplanted sail even on the long hauls, myth making was alive and well. The image of the 'glorious' last days of sail was created largely by retired seaman-writers as well as by authors of maritime novels and other voyage narratives who imbued seagoing with psychological and spiritual value."[13] So when *gadget* is used during this time, the word was meant to feel nostalgic.

Like many other works of nineteenth-century sea literature, *Spunyarn and Spindrift: A Sailor Boy's Log of a Voyage Out and Home in a China Tea-Clipper* (1886) is written in the form of a logbook. Toward the end of the novel, the protagonist, Tommy Davie, loans his log to other members of the crew to write in as a way to pass the time during the trip home. The sailors tell one another stories, recording in the book their "spun yarn." Today, when we describe telling a story as "spinning a yarn," we're actually drawing on the activity that sailors often paired with the sharing of stories. A ship's owners would buy scraps of old rope before a voyage that, during storms or downtime, sailors would work on as they sat together, pulling the junk rope apart into its constituent fibers and then "spinning" it, or loosely twisting two strands of the fiber together. Whether the yarn itself became a metaphor for the long, drawn-out tales sailors were so fond of or whether storytelling was simply the activity that happened around the spools is unclear.[14]

Eventually the log is passed to Baby, the greenest member of the crew. (It took a while for new sailors, leaving their farm jobs for work on ships, to lose their "green hands" and find their "sea legs."[15]) Baby shrugs off the invitation, claiming he has no idea what to write, but Davie urges him on: "Oh, yes you do! stick down anything, lies or not, it doesn't matter." Baby looks around for inspiration and spots the complicated network of ropes crisscrossing the ship's masts, sails, and deck. Lamenting how difficult it is to remember the name and use of all those ropes, Baby adds to the log,

> Down-hauls, and out-hauls, and brails, and braces, and halliards, and sheets, and tripping-lines, and clew-lines, and bunt-lines, and tacks, and lots more. If you happen to let go the wrong one there is such trouble. Then all the names of all the other things on board a ship! I don't know half of them yet; if the exact name of anything they want happens to slip from their memory, they call it a chicken-fixing, or a gadjet, or a gill-guy, or a timmey-noggy, or a wim-wom—just pro-tem., you know.[16]

It's a scene repeated in countless sea narratives: to learn the techniques of sailing is to speak an entirely new language.[17] The implicit joke here is that buntlines, braces, and downhauls are some of the most essential components of a ship's rigging; they are the lines that lower, raise, and direct the sails. It is absolutely imperative to be precise when naming each. "The unambiguous terms of the shipboard language of command translate into highly specific actions that, properly performed, forge a crew and ship into a single machine that harnesses the world's wind to the navigator's will," writes maritime literature scholar Mary K. Bercaw Edwards.[18] Baby, being green, simply replaces the language of ropework with an entire lexicon of object babble that is just as complex as the technical terms, yet entirely useless.

Ropework composed the most basic foundation of a sailor's expertise during the age of sail. Also known as marlinspike seamanship, ropework involved tying, maintaining, and rigging lines. A line is a rope that has been put to work. Again, Edwards: "Once a rope is rigged into its place, it is . . . always called a line. The term 'line' therefore refers to the function and its place within the context of the overall system of rigging. The material of which it is made is rope."[19] (Like words, ropes only acquire meaning in their use.) Ropework skills revolved around the quintessential sailor's tool, the marlinspike: a thin skewer of wood, metal, or bone used to work with marline, fray the ends of ropes to be spliced together, loosen knots too tight to untie by hand, and weave complex, multistrand knots.[20]

Of all the pro tempore words Baby used to describe the most important of maritime techniques—chicken-fixing, gadjet, gillguy, timmey-noggy, wim-wom—only one would eventually make it ashore. But by the time *gadget* began appearing in print to lend an authentic air to depictions of sailor life, the age of the tools it described was already over. Gadgets were quaint, romantic, retro. It's a joke that Baby is a character who doesn't even know what to do with a marlinspike, though his bunk was filled with "blocks, spunyarn, skeins of twine, marline, and so on"[21]—but of course, neither would the readers of this novel. For them too, it was just a gadget.

Today, nautical terminology serves as one of the richest sources of descriptive language at our disposal when navigating the internet.[22] Terms borrowed from manuscript and print technologies provide metaphors for the way we read on screens with almost one-to-one precision (scroll, bookmark, page, index, tab, bookshelf). But nautical terms apply to a wider array of digital tools and techniques: port (where ships, computers, and cables dock), log on (a practice taken from nautical logbooks), network (literally the work of nets—here, for ensnaring fish), surf (the crest of breaking waves or the internet), cybernetics (from the Greek for steering a ship), and pirates (shipbound and digital, both of whom Adrian Johns describes as global conveyers of diverse ideas whose reach rivals Enlightenment philosophers).[23] All of these terms have been transplanted from one area of practice to another. In the process, do the older techniques they once described—techniques for navigating and logging—hold sway over the use of our contemporary digital devices in any meaningful way?

Let's consider how the word *gadget* functioned in context among the spoken language of mid-nineteenth-century sailors. Decades before the craze for nostalgic nautical fiction, sailing ships carried diverse and multilingual crews that gathered new words every time they tied up at a new coast. That rich melting pot of languages settled into a common set of terms that was shared among sailors as a loosely knit community of practice. Individual ships also had their own unique terminologies, yarning, and forms of argument. Mary K. Bercaw Edwards distinguishes between the "occupational lingo" of the sailor profession as a whole and the "coterie speech" that was unique to the crew of a particular ship: the terms, myths, and stories that developed among its much more local culture. Sailor talk consisted not only of the technical terminology shared by all sailors but also the "occupational lore and . . . coded language that arises out of the shared experience of particular crews. . . . On any voyage the crew develops speech particular to that specific group of people, when the already arcane set of terms that is nautical terminology, the occupational dialect, shares into the more personal rubric of coterie speech."[24] (One can imagine the advent of wireless communication between ships substantially affecting these nineteenth-century bubbles of shipbound coterie speech.)

Gadget began somewhere within this economy of spoken slang. Given the traditional reliance placed by lexicography (the writing of dictionaries) on textual evidence, the actual origins of the word in speech remain hazy. But it was occasionally captured in print texts written by and for sailors, who had a surprisingly rich textual life. Literary historian Hester Blum notes that "seamen were literate to a degree unusual among laboring classes—the best estimates place sailor literacy in a range from 75 percent to as high as 85-90 percent—and they engaged in a lively system of exchange of books and other reading materials among ships."[25] A sailor in the 1840s could have picked up books at dockside newsstands, browsed the printed catalogs of shipboard libraries, settled down with a serial novel in portside reading rooms, and traded books with fellow sailors. This was an entirely different textual network than the one cast by the amateur antiquarians communicating in *Notes and Queries*.

Many of the novels in question were written by sailors for an audience of other sailors and contained actual advice on techniques of the trade—what Blum calls "an aesthetic of mechanical precision."[26] When the novels gestured to a potentially broader readership, they didn't shy away from the complexity and richness of coterie speech. Richard Henry Dana Jr., who first worked as a merchant seaman in 1834, described his popular memoir, *Two Years before the Mast* (1840), as an unmediated account of his experience "written out from a journal which I kept at the time, and from notes which I made of most of the events as they happened. . . . In preparing this narrative I have carefully avoided incorporating into it any impressions but those made upon me by the events as they occurred."[27]

In the memoir's preface, Dana prepares his readers for the nautical terminology and technical detail they are about to encounter, arguing that it can actually benefit readers if they come to nautical fiction without any knowledge of these tools, so long as the technical details gradually unfold throughout the course of telling a story:

> There may be in some parts a good deal that is unintelligible to the general reader; but I have found from my own experience, and from what I have heard from others, that plain matters of fact in relation to customs and habits of life new to us, and descriptions

of life under new aspects, act upon the inexperienced through the imagination, so that we are hardly aware of our want of technical knowledge. Thousands read the escape of the American frigate through the British Channel in [James Fenimore Cooper's] "The Pilot," and the chase and wreck of the British trader to "The Red Rover," and follow the minute nautical manoeuvres with breathless interest, who do not know the name of a rope in the ship; and perhaps with none the less admiration and enthusiasm for their want of acquaintance with the professional detail.[28]

Like Baby in *Spunyarn and Spindrift,* Dana knows that his reader will stumble through this story world without even "know[ing] the name of a rope in the ship." And yet the "professional detail" of things like halliards, tacks, and brails becomes legible for nonspecialist readers in the process of reading a story of sailor customs and habits. In *Two Years before the Mast,* marlinspikes are everywhere: floating on a lanyard worn around the neck of a sailor who went overboard, barely grasped in the freezing hands of another. Dana doesn't stop to explain them or other shipboard tools; he simply lets them motivate the narrative action. The specific properties and functions of each tool fall away, leaving behind a bunch of gadgets that provide an aura of nautical authenticity as well as the sense that all of these things must do *something* important. Today, we might think of this literary technique as a close cousin to technobabble, which science fiction would refine as one of its defining characteristics later in the century. Science fiction authors invent new names (neology) for all of the new things (novums) gradually woven throughout the course of the narrative as part of its world-building fabric.[29] In the case of nautical fiction, these names and things, new to the reader, have real-world referents.

As an alternatively functional and fictional device, the gadget is a useful object for conceptualizing the relationship between a tool and an individual user's imagination of how that tool works. In this sense, *gadget* from its earliest use in print didn't just refer to particular tools whose names couldn't be remembered. The word was already fully imbricated in an entire range of connotations:

in this case, it was a sign of what someone knew, and the signature of a broader public's hazy sense of what those gadgets did in the first place, how they felt in the hand, what they were capable of. The deceivingly simple nautical origins of the gadget as a generic name for anything in fact already contained some of the complexities that would provide the word such wide applicability throughout the twentieth century. Although we can rely on *gadget* and other nautical terms as metaphors to describe complex technologies, the way we do things is tied up in the communities in which we do them. Once secondary representations of those techniques begin to circulate, there is so much that is lost in translation.

While readers of *Spunyarn* reminisced over bygone sailing techniques, a team of word hunters outside London culled through mountains of paper slips mailed from across the country. The Philological Society, another group of hobbyist antiquarians, first hatched the plan for a massive crowdsourcing project during a meeting in 1860. The idea was to root out the origins and evolution of every word ever used in the English language. Though far more systematic in its scope than the occasional curiosities discussed in the forums of *Notes and Queries*, this endeavor too was nevertheless the product of inexpert techniques.

It worked like this. Lexicographers in Mill Hill, a small suburb of England, mailed thousands of books of all kinds to volunteer readers across the country, along with a request to read them with an eye toward interesting words: "Make a quotation for every word that strikes you as rare, obsolete, old-fashioned, new, peculiar or used in a particular way. Take special note of passages which show or imply that a word is either new and tentative, or needing explanation as obsolete and archaic, and which thus help to fix the date of its introduction or disuse."[30] This army of volunteers spanning the English-speaking world was, for the most part, anonymous. They were unknown to the *Oxford English Dictionary*'s editors and to one another as they took note of interesting words, working from home. As Lindsay Rose Russell has argued, "the dictionary's entry into the home both prompted and normalized women's participations in lexicography." In fact, Russell identified no fewer than 250 women who were among the most active of these volunteers, with several of them individually "submitt[ing] more than 10,000 citations from their individual reading."[31]

At some point around 1880, a volunteer correspondent would have come across the word *gadget* in her reading and decided it to be noteworthy. She wrote the word on the top left corner of a half sheet of paper, known as a quotation slip, and then copied the passage containing the word, finally noting its source, author, and date. (It became a problem that correspondents tended to hunt for unusual words while skipping over the more pedestrian ones. Common words were the most in need of data—words like *set*, which proved to be the most complex in the English language.[32]) Once the correspondent had written up a number of these slips, she mailed them in a bundle back to the lexicographers. Mail carriers delivered sacks of these slips to Mill Hill, with about a thousand slips arriving daily. The slips were then pored over and sorted by arrangers and subeditors like Ellen Skipper and a woman with the surname of Scott. Skipper and Scott unpacked the slips from their heaping piles and packages, arranged their headwords into alphabetical order, and slotted the paper into a grid of pigeonholes.[33] Over the years, some two million slips were collected.[34]

The production of the *Oxford English Dictionary*, as it would eventually be called, took sixty-eight years to complete. Often cited as the ground truth on deep dives into any word's origins, this landmark of language scholarship began as the product of a community of amateurs and volunteers. At a time before the sciences of language (like linguistics and phonology) had solidified, the makers of the dictionary essentially invented their research methods as they went along. For Anatoly Liberman, an etymologist who currently works for the *Oxford English Dictionary*, in many ways this remains the case today: "Everything in etymology is conjecture and reconstruction."[35] Their methods read like detective work across intersecting axes of time and space. There's antedating, or the slow hunting through the written record to find ever-earlier instances of a word in use in order to get as close as possible to the source of its historical emergence. A word's ancestor is its etymon; it usually has a completely different meaning than its modern sense, its center of gravity shifting with time. Then there's orthography, or assessing differing conventions in spelling, capitalization, and punctuation across many languages in order to identify where that word may have come from; it's like viral contact tracing for words across the globe.

In 1880, the work was proceeding in alphabetic order. Some letters were subjected to longer delays than others, depending on how organized the subeditor assigned to a given letter was. According to historian Simon Winchester, "some of the sub-editors had put their hundredweight collections of papers into hessian sacks, and then left them to rot. . . . A dead rat [was found] in one of these, and then in another a live mouse with her family, all of the creatures contentedly nibbling away at the paper, making bedding for themselves out of years worth of lexical scholarship."[36] The slips containing examples of gadgets were given to J. Bartlett, the subeditor in charge of all words from *G* to *glasscloth*. Along with Eleanor Spencer Bradley and A. M. Turner (women who were assigned to the letter *G* as paid staff editors), Bartlett compiled these words from 1888 to 1891, and then again in 1897 and 1898, in order to incorporate new words that had been received over the past decade.[37] As collaborators on the *Oxford English Dictionary* gradually accumulated these time-stamped examples of *gadget*'s appearance in print, a portrait of the word's evolution began to develop, allowing Bartlett, Bradley, and Turner to finally write the word's definition.

In computational text analysis today, we might think of the work done by Skipper and Scott—the daughters of local artisans who were some of the first women on the payroll of the *OED*[38]—as a form of tokenization, stemming, and lemmatization. These are the steps taken to preprocess a text for study (or, in the case of the *OED*, an entire language) by breaking it down into its constituent parts. Rather than preparing a text with algorithms for what's called natural language processing (NLP), the nineteenth-century work proceeded by relying first on scores of volunteer readers to parse thousands of books into their elementary parts, then on human computers like Skipper and Scott to compile those words. Preprocessing looks similar across lexicography and NLP, but the resemblance ends there; in fact, one of the foundational articles for modern NLP critiques dictionaries as an objective source for the meaning of words.

Writing in 1957, linguist John Firth described the citational practice of lexicographers as "literary" and "arbitrary," recommending instead a statistical approach that measured the meaning of a word by counting the other words most frequently occurring alongside it. "You shall know a word by the company it keeps!" he famously wrote.[39] Firth and mid-twentieth-century linguists like Margaret Masterman—a student of Wittgenstein's who went on to become a key figure in the early days of computational semantics and machine translation—took their inspiration from thesauri rather than dictionaries.[40] Lydia H. Liu distinguishes between the two approaches: "If the *OED* is philological and linguistic, the thesaurus is philosophical and mathematical where the meaning of a word is not defined by other words so much as determined by what surrounds its use or by the location where the word occurs (its context) with a certain pattern or regularity."[41] It's well known that Firth's "know a word by the company it keeps"—now referred to as the distributional hypothesis—serves as the theoretical underpinning of NLP and generative text technologies. Less well known is the fact that the paragraph containing that line begins with Wittgenstein's "meaning of a word is its use in the language," as if the two insights are one and the same.[42]

Whether we rely on dictionaries or thesauri, notecards or computer punch cards, no method can fully recover the embodied, three-dimensional world in which *gadget* was embedded. The *OED* didn't produce definite answers, just stories relayed through dank basements and dusty pigeonholes. When it comes to the nautical lexicon, people like Skipper, Scott, Bartlett, Bradley, and Turner were counting fish in the sea. And for all the sense that NLP-driven "artificial intelligence" tools seem to make, just because they identify patterns in the use of words doesn't mean they have a workable model of the world in which those words were applied.[43] The sensory environment that produced *gadget* remains nearly impossible to grasp: "the gear and fittings and architecture of the ship. . . . rails and ropes, masts and booms and barrels and brooms . . . the sound of hasps honing the steel of lances and harpoons [that] filled our ears throughout the day."[44] This is what I take Wittgenstein's "meaning in use" to imply. A word is embedded and embodied in the three-dimensional, social contexts in which it is used, not just in relation to the other words around it. Like a rope becoming a line when it is put to use, *gadget* only takes on meaning in context.

According to the *Oxford English Dictionary* correspondents, *gadget* had become common in the spoken lexicon of nautical expertise by 1870, with anecdotal evidence suggesting that it may have been used as early as the 1850s.[45] One volunteer reader antedated the word with its appearance in *Spunyarn and Spindrift*, the earliest textual evidence found. Beyond that, there are no true or even accurate answers as to where *gadget* came from, only approximations and stories: a dialogue between technicians and lexicographers, between craft and its description.

Nineteenth-century sources sometimes spell the word *gaget* or *gadjet*, although it appears to have emerged entirely independent of homonymous babble like *gewgaw, gaud, gambol,* and other descriptors of various objects. One apocryphal tale attributes the word's origins to miniature figurines of the Statue of Liberty made by Gaget, Gauthier, & Co., the Parisian metalworkers who constructed the statue's copper shell in the 1880s. According to this origin story, the manufacturer's name was printed on the base of the figurines and became a shorthand for the souvenirs themselves. Street sellers walking around Liberty Island would ask passersby, "Do you have your Gaget?"[46]

Another story suggests a correlation with the French *gâchette*, a word that sends us down a rabbit hole of terms referring to mechanisms—hinge, toggle, trigger—each of which have had their own metaphorical afterlives. (We toggle between two topics, everything hinges on this, it triggers a response.) *Gâchette* refers to a component within a complex mechanism, especially the trigger of a gun. The word is a diminutive of *gâche*, which is part of a latch. The object itself couldn't be more banal, but its descriptions are incredibly complex. Think of the latch that locks a creaky screen door or a nice cupboard; the *gâche* is known in English as the catchpiece, staple, or ring into which the toggle or hook fastens. Some of these words remain in perennial use as they spread to figurative ground, while the physical objects they originally referred to retreat into the background. Meanwhile, those words that found nowhere new to grow—think *catchpiece*—have seemed to wither away.

A range of other French correlatives to *gadget* have emerged: *gagée*, dialect for a small tool; *engager*, meaning to engage one thing with another; and *gauge*, an Anglo-Norman French word that first entered the Scottish lexicon as *gadge* with *-et* as the diminutive suffix.[47] A 1909 French-German-English translator's dictionary of technical, industrial, and commercial terms as well as a 1918 aviator's pocket dictionary both translate the English word *gadget* into French as *dispositif*, a term that would become central to discussions in film theory about the ways media technologies, or apparatuses, have the power to shape social discourse.[48] Today, the French phrase *c'est du gadget* refers to ideas or objects that are trivial or irrelevant to the subject at hand.

We could accumulate countless other origin stories, like the two characters who pen the famously enigmatic opening sections of Herman Melville's *Moby-Dick*. "A late consumptive usher to a grammar school" provides a complete etymology of the word *whale*, "dusting his old lexicons and grammars" to root out the historical meanings attributed to the animal. A sub-sub-librarian, "a painstaking burrower and grubworm of a poor devil of a Sub-Sub," shares ten pages' worth of "whatever random allusions to whales he could anyways find in any book whatsoever, sacred or profane."[49] But much like lexicographers and sub-sub-librarians, we who inhabit a future that has outlived the world of commercial sail remain outsiders to the tools and techniques then associated with the word *gadget*.

Technical terms have a peculiar half-life. They age differently than other words as the tools and techniques to which they refer evolve or when different communities of practice apply them in new ways. It was for this reason that Samuel Johnson banished the word *marline* from his 1755 *Dictionary*, the preeminent English dictionary until the *OED* came along, because "sailing is one of the arts not liberal or confined to few."[50] Despite its branching variations and broad applicability (the noun sprouted verbs—one could *marl* or *marl down* by winding a knot with marline to prevent it from unwinding), *marling* remained a kind of forbidden word.

In reading the language of practitioners, we encounter the strange sense of words that now refer to something entirely different, or the insular use of technical terms that mean nothing outside the bounded world of craft and its applications. The grammar of nautical technique reads today almost like incantations or glamour (a word that itself is a corrupted derivation from the word *grammar:* language that has been weaponized, diseased, twisted into a spell). Nautical handbooks of the nineteenth century contain the most beautiful descriptions of rope technique, listing thousands of numbered instructions with economical prose that feels like it contains more prepositions than proper nouns. Here is an 1860 description of how to tie the French, or "Single Shroud," knot:

> Crown backwards, lefthanded, the strands of each end; then dip the ends that lie *from* you to the left of those that fall down *towards* you: haul them into their places.[51]

And a description of a magic trick using cat's cradle figures woven between the fingers:

Hold all taut and cast off the loop from the right forefinger and at once pull out the loop that remains on the second finger. All complications will at once vanish.[52]

Over time, even the driest technical descriptions are hollowed out of their concrete meaning, leaving behind only the filigreed outlines of poetic abstraction. Techniques of the 2020s will acquire a similar patina with time. We live in yet another moment when little feels unchanged save for the clouds and our bodies beneath them, as Benjamin put it. As I write this, language struggles to keep pace with breathtaking advances in the overlapping fields of NLP, machine learning, and artificial intelligence. We alight on names for new techniques like *prompt engineering* and new descriptors for text like *synthetic, automated,* or *generative* to wrap our heads around systems that are inscrutable even to their creators. Soon after the November 2022 release of ChatGPT, for example, I heard a radio advertisement for an AI company promising that its product was "hallucination-free, LLM-agnostic, IP-compliant," a phrase that would have been nonsense to the listening public a mere six months earlier. It's not that we're at a loss for words to describe new ways of doing things. It's that we're triangulating a three-body problem as rapidly evolving tools, the names given to those tools, and the tacit knowledge bound up in their use all move in parallax with respect to one another.

What makes *gadget* unique as a term of art is its indeterminacy: sailors devised the word to mean anything at all, whatever you happened to be pointing at. The word is a placeholder, an empty container. This wide applicability allowed the word to spread from a specialized community of practice to other technical lexicons and eventually to general, everyday use. Vita Sackville-West draws on exactly this sense when she writes, "What an odd little word 'gadget' is, almost a gadget in itself, so small and useful." Scouring her toolshed for just the right implement, this poet, novelist, and garden designer sorts through "the walking-stick shaped like a golf-club with a cutting edge to slash down thistles . . . the little wheel on the long handle, like a child's toy, which you push before you and which twinkles round, cutting the verge of the grass as it goes." Whatever she picks up, *gadget* fits almost "any small tool, contrivance, or piece of mechanism not dignified by any specific name."[53]

Though *gadget* is an empty container, the shape of that container transforms into distinct shapes from the nineteenth to the twenty-first centuries, lending the word different connotations in different moments. The way *gadget*'s center of gravity shifted among sailors at the turn of the twentieth century—from ropes and pulleys to cogs and gears—as sail was replaced by steam provides a microcosm of an ongoing process in which we still take part. By the time Sackville-West wrote her toolshed essay in 1940, *gadget* had begun to imply cheaply made tools with multiple functions that "seldom work. . . . Different from gadgets are the time-honoured tools which hang in the dusty brown twilight of the tool-shed when their day's work is done."[54] *Gadget* would come to signify other genres of tool and ideas about those tools throughout the twentieth century, from fashionable accessories in the 1930s to consumer electronics in the 1980s, from atom bombs in the 1940s to novelty items, cheap gimmicks, magic tricks, and other toys in the 1950s. At some moments, gadgets are handmade; at others, they are mass produced. Gadgets have been important and trivial, obvious and clever; they have named an entire tool and a specific component within it. The connotations of the word itself have spread from tools like can openers and smartphones to ideas about tools like automation and triviality, from people like aviators and carpenters to habits like cutting corners and distraction.

We can read this keyword in context, and with just a few skips across the surface of the twentieth century, we can see how its semantic drift captures the evolution of a discourse on technology. Roughly every decade, the word lands on a distinct set of tools, techniques, and ideas: a *gadget* is

any "nameless, improvised thing" (1908),
an "accessory ... difficult to remember or define" (1919), and
"anything you may be pleased to imagine" (1919).

It is "something which seems to be useful but isn't" (1937),
"anything which, designed to simplify life, actually complicates it" (1940),
"a device for doing something that nobody knew needed doing until a gadget was invented to do it" (1949).

Gadgets "obviously have had no place in the constitutional scheme of advancing scientific knowledge" (1950),
they are "mechanical improvements which should have been obvious to a person having ordinary skill in the art" (1952),
and the gadget "differs from the complex machine in that it minimizes the importance of the human being whereas the ordinary machine augments a person's capacity to do work" (1959).

To Jack Kerouac, gadgets require a "delicacy of thumb and forefinger" (1952),
to John Updike they are "dainty Swiss" things (1966),
to Herbert Marcuse they "keep [individuals] occupied and divert their attention from the real issue" (1968),
and to Don DeLillo they are featured in "degenerate ceremonies ... that mock our bodies" (1976).

The gadget is "the home computer" (1989)
or "the caller ID box" (1991).
They are airport "gates of electronic surveillance" (1977),
"a new type of modem" (1988),
and "small programs that you can add to your computer desktop" (2009).

As Marshall McLuhan put it in 1951, "gadgets and gimmicks did not begin as physical objects, nor are they only to be understood as such today."[55]

The history of this word exerts a pull on the way it is used today. As you, reader, consider the crushed look of a friend whose expensive new smart speaker you just called a gadget, you are engaging in a dialogue with the cumulative history of that term's application. Even at our remove from the lived experience of historical and highly specialized technical practices, we participate in this dialogue through the way we use *gadget*, both the word itself and the tools we apply it to. Because of this semantic drift, it's a word whose modern sense barely echoes some of its historical meanings, like the words *egregious* (wonderful, extraordinary, remarkably good) and *actually* (something characterized by action and deeds).[56] The shifting sense of *gadget* provides us with distinct portraits of how people thought about their tools during particular moments in history. To call something a gadget is to render a judgment on what that tool does, how it works, and what it means in your life. These judgments are shaped by historical context, influenced by technological form, and—most importantly—rendered by users themselves.

Clustering these atomized associations together allows us to find historical inflection points where *gadget* began to mean something new on a collective scale at particular moments. Up until 1950, for example, *gadget* is applied to either inner components or accessories added on to electrical devices, like vacuum tubes or frequency meters for ham radio sets.[57] This usage would disappear by the end of the 1950s as small, neatly printed circuit boards developed during World War II through the Auto-Sembly process ended up replacing the mess of spaghetti wires filling the guts of consumer electronics. This meant that repairs previously performed by users with a soldering iron or even a quick shake of loose connections became impossible. This movement, in which the visibility and tinkerability of components diminished, was accelerated by the replacement of the vacuum tube with the tiny transistor, which was first developed in 1947 and available to consumers beginning in 1950. As the price of transistors fell and manufacturing techniques improved, they allowed for a wave of increasingly small consumer electronics.[58] It's around this time that *gadget* begins referring to the entirety of a device rather than its inner components or attachments.

The discourse of gadgetry followed the scale of user interactions with electronic equipment as it panned out from known, internal components to the entirety of an increasingly sophisticated piece of consumer technology.[59] Somewhere around 1955, users of technology in America began to think more about the entirety of their tools than they thought about the parts and components those tools contained. For Donna Haraway, this trend toward miniaturization and virtualization led, by 1985, to a form of cyborg consciousness:

> Contrast the TV sets of the 1950s or the news cameras of the 1970s with the TV wristbands or hand-sized video cameras now advertised. Our best machines are made of sunshine; they are all light and clean because they are nothing but signals, electromagnetic waves, a section of a spectrum, and these machines are eminently portable, mobile—a matter of immense human pain in Detroit and Singapore. People are nowhere near so fluid, being both material and opaque. Cyborgs are ether, quintessence.[60]

While Haraway saw in the cyborg the potential for new political movements and new forms of transgression, the late-twentieth-century shift from physical device with known internal components to magical, miniature gadgets promising cyborg consciousness was not evenly distributed. As Samuel R. Delany wrote in 1994, "The black boxes of modern street technology (or the white boxes of computer technology—not an accidental distinction, I'm sure) put us in a very different relationship with the inner workings." On the one hand, Delany saw young Black Americans carrying "matte-black, plastic boxes . . . from the beepers, the Walkmen, the Diskmen, through the biggest ghetto blaster" and read those objects reflecting "a technological culture that's almost entirely on the receiving end of a river of 'stuff.'" The iconography of high technology, on the other hand, revolved around white desktop computers and a "class of people who know the incredibly complex spells and incantations needed to get the stuff to work."[61] In any given period, there is a "sign language" to the types of tools people use, as Delany described it. By the 1990s, the more likely that a tool was to be seen as a gadget, the less likely the user was expected to have any agency over that tool's design and construction.

Today, gadgets have evolved from clever little trivialities or handy multitools into platforms that influence our politics, attention, relationships, and memories. In the process, a tool that was mapped onto gender (handymen; boys and their toys) now instead seems a marker of generation (screen-addicted millennials, who studies show actually use their phones less frequently than Generation Xers).[62] And while the word refers throughout its history to specific objects, it also implies particular techniques or styles of doing things.

Technique encompasses a complex dialectic between tools, their uses, and the words describing both. Whether we are maintainers of ropework or adepts of texting in T9, every day we apply techniques that are in the process of undergoing dynamic change along multiple axes. We all know that tools and paradigms of media technologies evolve: radio becomes television becomes internet, letter writing becomes emailing becomes texting. As those media and their associated tools change, so too do the words describing them. Think of the gradual succession of prefixes to the word *phone* in American English: car (1980s), mobile (1990s), cell (2000s), and smart (2010s). Less widely understood are the ways that broad patterns of use change over much longer timescales. Even though the word *marl* was long ago banished from the dictionary, today, centuries since the marline to which it referred has faded into insuetude, the underlying technique can be found on satellites orbiting the earth, "whose wires are wrapped together with small strands of string using the ancient 'marling down' technique." The technique can be seen "binding the fiber-optic cables in the server facilities that power the Internet."[63] How many other mechanical or analog techniques live on in the age of tech?

In the end, it feels impossible to separate the three categories of tool, use, and description. Descriptions of the marlinspike often hover between a specific tool and a broader way of going about a task. Harold Calahan, in a nautical manual titled *Gadgets and Wrinkles*, puzzles over whether the marlinspike belongs in the category of gadgets or wrinkles. While "a gadget is a machine, an invention, a mechanical means of achieving a result, a wrinkle is a method of procedure. . . . To the seaman, a gadget is a thing, and a wrinkle is a method, and both of them for the most part unusual and unstandardized." The problem that Calahan encounters throughout his handbook is that a clear distinction is not so easily made between a gadget and a wrinkle, between a tool and a technique. He continues:

> Take a familiar example. You are about to tie two lines together. You tie them into a weaver's knot. Standard practice so far—no gadgets, no wrinkles. Then realizing that there is going to be a terrific strain on that line and that the knot will be pulled so tight that you will never be able to untie it again, you decide to slip a toggle into the knot. That's a happy thought, for you can always take a hammer and drive out the toggle and the knot will be loose. What is that toggle—a winkle or a gadget? [. . .] The line of demarcation is too indistinct.[64]

The marlinspike that forms a toggle to easily undo tight knots in one context could also be used as a handle to tighten knots in another. In the marline hitch, the marlinspike can be rotated by hand to secure large parcels or chests, or to lash objects like hammocks to the ship's rigging. In these scenarios, it becomes impossible for Calahan to say exactly where tools end and techniques begin. It's a problem that, over and over, produces these long digressions in a manual that's supposed to be as functional as possible. (Imagine someone on deck pulling out a handbook and trying to figure out how to undo a knot by desperately thumbing through page after page of philosophical meditations!)

From its earliest beginnings, then, *gadget* was a term that signified particular tools as well as the techniques that emerged to solve problems: it's not only the bucket in which a rope will keep from tangling when dropped, but the particular technique of twisting the rope "clockwise between the forefinger and the thumb of the right hand" as it's fed in.[65] This may seem like a strange concept in English, but for Continental languages, there is little terminological distinction between the tool and the technique. Gadgets extended from tools to the stylistic flair cultivated by particular sailors who prized solving problems in the most economical way possible, manifesting brilliance in the materials. A single tool could solve countless problems, and the genius was to be located in the body of its user. Take the *Pequod*'s carpenter in *Moby-Dick*, the crew member who was "singularly efficient in those thousand nameless mechanical emergencies continually recurring in a large ship." Ishmael compares the carpenter with probably the quintessential gadget to be found in *Moby-Dick*, a multitool then known as a carriage knife:

> He was a pure manipulator; his brain, if he had ever had one, must have early oozed along into the muscles of his fingers. He was like one of those unreasoning but still highly useful *multum in parvo*, Sheffield contrivances, assuming the exterior—though a little swelled—of a common pocket knife; but containing, not only blades of various sizes, but also screw-drivers, corkscrews, tweezers, awls, pens, rulers, nail-filers, counter-sinkers. So, if his superiors wanted to use the carpenter for a screw-driver, all they had to do was to open that part of him, and the screw was fast; or if for tweezers, take him up by the legs, and there they were.[66]

The manual intelligence of the *Pequod*'s carpenter is a trait that Mauss attributes to all humans—"man is an animal who thinks with his fingers"—and not just the carpenters among us.

But I often wonder what will be retained about the way we use digital devices today in the words that are left behind, especially given that there is so little language provided to users of these devices. Think about the last gadget you bought. Did it come with an instruction manual? Maybe, at most, a quick-start guide, but more likely than not, the device you purchased was designed to feature user interactions that are intuitive, frictionless, magic. This is one of the signature ironies of user experience design today: the more complex the engineering of a device becomes, the less instruction users need in how to use it. Many of the words we rely on when using these devices are inherited from older techniques that—like *hinge* and *toggle*—have migrated from their tangible referents into metaphorical dimensions. Even with the embarrassment of data constantly generated by digital devices and their users, there are things about the texture of using them that will be near impossible for future historians to recover. During an era of constant surveillance of our most minute habits, it seems strange to consider that anything will actually go unremarked or unrecorded. But looking to past gadgets provides indications of the shape of things that will in fact be forgotten.

From Michael Penn, *Tune In* (2017). Book designed by Natasha Hulme.

4

The Custody of Automatism

I'M SITTING IN A CIRCLE WITH A GROUP OF NEIGHBORS, librarians, technologists, and organizers. Spools of Ethernet cable lie at our feet. Following the instructor's lead, we each cut a portion of the cable, strip its plastic outer casing, and peel back a thin shield of foil to expose four twisted pairs of candy-colored wires. The key to untangling each tiny pair, the instructor tells us, is to pinch the base in one hand and pull them taut with the other, dragging our fingers along their length until we've separated the individual strands. (For some reason, I always have trouble doing things this way and just end up unwinding my wires like bread bag twist ties.) We arrange the eight strands into the proper order—their pinout—and slide them into a plastic cap before finally using a viselike tool to crimp the cap down onto the wire. Like sailors gathered to spin a yarn from junk rope on a long voyage, we now have several made-to-size Ethernet cables ready to be plugged in.

Cable crimping is the first step in a training session for volunteers with Philly Community Wireless (PCW), a cooperative mesh network and digital equity organization devoted to empowering residents of North Philadelphia to grow and maintain their own connections to the internet.[1] Today, volunteers and seasoned technicians with PCW are going to hook up a local homeowner who would like to broadcast a free Wi-Fi connection to anyone enjoying the benches and community gardens in an adjacent park. We take our fresh-cut cables outside, fanning out around the base of a brick rowhouse. Those of us with the stomach for it squeeze out windows or scamper up ladders to reach the rooftop. One technician cradles a laptop, ready to read information from a Wi-Fi antenna that she bolted into place on an old TV antenna. She tosses a bundle of cable down the building's side to a team waiting below, ready with a plaster drill to bore through century-old brick and feed the cable through to the basement, where it will be connected to a router. After years of trial and error working in the neighborhoods of Fairhill and Kensington, we have learned to square up the drill when applying pressure, to use particular bits depending on the wall's material, to keep the drill's hammer function off, and to feed the cable through the brick with the help of an electrician's tool called fish tape.

At its heart, the idea behind mesh networks is to share a single connection to the internet through a web of interchangeable access points. Greta Byrum, codirector of the digital equity laboratory at the New School, explains: "Instead of requiring a centralized hub to direct network traffic, a mesh operating system automatically searches for the best path for data to travel. Devices (even computers or phones) can become 'nodes' or connecting points that enable data to hop from place to place until it reaches its destination. If a node fails or breaks, the network automatically routes around it through other nodes."[2] Philly Community Wireless formed during the Covid-19 pandemic with hopes of using mesh networks to connect people who still lacked access to high-speed internet.[3] When schools closed at the beginning of the pandemic, roughly half of Philadelphia's kindergarten through twelfth-grade students lacked an internet connection at home, rendering the Chromebook laptop lending program offered by the school district useless.[4] According to a study released in the pandemic's second month, 80 percent of white residents had broadband subscriptions, compared to only 53 percent of Black and 44 percent of Latinx residents in the Philadelphia–Camden–Wilmington metropolitan area.[5] At that point, internet accessibility in Philadelphia corresponded so closely with race that a map of broadband connection rates by neighborhood was essentially identical to a map of the city's whitest neighborhoods.[6]

While groups in cities around the United States found a similar dynamic at play, mesh projects from Oakland to Baltimore were able to hit the ground running thanks to influential curricula developed by the Detroit Community Technology Project, founded by Diana Nucera.[7] She defines community technology as "a principled approach to technology that is grounded in the struggle for a more just digital ecosystem, placing value on equity, participation, common ownership, and sustainability."[8] From the perspective of community technology, building mesh networks means not only maintaining the code base and build-out, but also "fostering relationships of trust and cooperation among neighbors, who must work together to make decisions about network design, services, access protocols, security, and long-term sustainability."[9] Rather than parachuting in shiny new hardware, mesh network groups and community technologists host participatory design workshops and technical trainings that empower communities to maintain and grow their network connections in relationships of care.[10]

The pandemic foregrounded the urgency of finally addressing digital inclusion in the United States a decade after the United Nations declared internet access to be a fundamental human right.[11] But in the absence of a competent policy response at either the federal or local levels to ensure access to a medium now understood to be almost as fundamental to human flourishing as water, communities have devised various forms of making do to fill the gaps. Across the country, groups like PCW organized to build networked computing infrastructures from scratch, using combinations of hardware, software, policy stopgaps, and community ownership.[12] In theory, we would use an old medium of the 1920s (radio waves) to ensure access to a new medium of the 2020s (the internet). In practice, a group of people who began with an interest in wireless networks slowly realized they would need to learn about ladder safety, consensus making, masonry, and cooperative ownership models.

My hands coated in brick dust provided me with an uncommonly visceral experience of technique in an era of smart devices and ubiquitous connectivity; typically, internet infrastructures recede into the background. For those of us privileged enough to afford high-speed internet access, this stuff is all just supposed to work. But today, the techniques we shared for braiding and splicing have made a direct impact on the way we connect with one another around the park. If just one strand of the inner wires of our crimped cables is out of place, then the signal moving through our network might be slowed or corrupted. Media artist Mimi Ọnụọha calls these tightly braided cables "the veins through which digital information travels."[13] In her series of installations and videos *The Hair in the Cable,* Ọnụọha and others gather to weave cables together with spices, cloth, and hair, drawing connections between modern infrastructures and traditional practices from Igbo culture. In Philly, as we build out and maintain this network, it feels amazing to know that techniques spanning centuries of material know-how are just as important as digital skills like network engineering. Everyone has some expertise or labor to contribute, physical or cognitive, based on what each person knows: how to facilitate and document meetings, how to adopt a new device to the network, how to get the word out through tabling and fliers, how to hold the rungs rather than the sides of a ladder.

Community technology runs counter to the prevailing winds of technique in the 2020s. Rather than atomized users accessing invisible networks guided through algorithmic suggestion, the work behind groups like PCW is collaborative, manual, visible. The challenge of this work is to grow locally minded communities that practice conscientious technique at a time when the way we use digital devices has become especially adversarial. Our inscrutable gadgets homogenize and track us; their algorithmic filters are increasingly the only window we have on the surrounding world. Chapter 1 of this book describes the eerie presence of these devices by exploring how technique encompasses both tool and a user's imagination of what that tool does. Chapters 2 and 3 paint complementary portraits of technique from the perspectives of philosophers and technicians, both of whom see meaning in tools only when they're put to use. Now I would like to close my story by asking whether the terms set forth up until now—terms equally applicable to ladder, cable, knot, plow, radio—are still useful for an era of distraction, automation, and surveillance. What becomes of technique when the tools we now use are designed to record, predict, and influence the ways we use them? What will the automation of everyday interactions through so-called AI mean for the way we do things? In this chapter, I explore some of the representative forms of technique today—forms that pose genuinely unique challenges to our cultural and political moment—and some of the more hopeful historical through lines that I see potentially informing those techniques.

Community technologists are building cooperative infrastructures for networked communication at a time when the broad public discourse on tech has become especially fraught. A mere short list of foremost concerns ranges from data privacy and electronic waste to misinformation, smartphone addiction, and massive job automation through AI. But while community technology proposes techniques for rethinking the function, ownership, and governance of digital media, the broad public discourse on the tech industry's excesses have remained largely focused on individual user experience.

One such response is digital minimalism, a movement that has emerged to cope with ever-intensifying attacks on the ability to maintain an uninterrupted train of thought in the age of tech.

Self-described digital minimalists seek to reinforce the permeable membrane of their attention through screen detoxes, software settings, productivity apps, and other forms of tech fixes. Approaches within the movement fall into a few different categories. First are the books that seek to identify and change the habits that lead users to constantly seek distraction and refuge in their phones. Proponents in this camp offer prompts like "spend time alone" and "hold conversation hours."[14] A related group consists of tech industry employees and insiders who have seen the light and devoted themselves to fixing the social harms their algorithms have created. Former Google engineer Tristan Harris, for example, founded the nonprofit Center for Humane Technology to draw attention to the ways recommendation algorithms manipulate our cognitive biases in order to keep us scrolling. The group has begun a laudable campaign to "reverse the digital attention crisis" by pressuring tech companies to pivot from business models built on cognitive exploitation and instead "create market conditions for humane tech," albeit without a real road map for doing so.[15]

A second category of digital minimalists tries to solve surveillance capitalism with tech fixes. For their part, tech companies now allow users to seemingly calibrate their attention through options recently added to the system preferences of all major mobile platforms. These features are known as Digital Wellbeing in Android, Screen Time in iOS, and Zen Mode on OxygenOS. A digital minimalist might be further outfitted with gadgets like IRL Glasses with polarized lenses that block images on screens, and slabs of solid plastic called NoPhones that act as pacifiers for smartphone addicts. Digital minimalists can also choose from the Google Paper Phone, a paper printout folded like a zine containing all of the information you'll need for the day, and the Light Phone, an app-less "dumb phone" with an e-ink display that only supports calling and texting.

A third group of digital minimalists draws from the medicalized discourse of screen detoxes and tech fixes used by both groups above. But they go one step further by attributing the twenty-first-century crisis of attention not only to overreach by the tech industry and its powerful data models but also to a certain moral failure on the part of users. Public philosophers James Williams and Matthew Crawford both describe attention as a battleground for human agency. Both also seem almost nostalgic for an earlier era when humanity could rely on an "off-the-shelf package of religious and cultural constraints" to regulate our attention, agency, and values. Williams writes, "In the twentieth century the rise of secularism and modernism in the West occasioned the collapse—if not the jettisoning—of many of these off-the-shelf packages of constraints in the cause of the liberation of the individual." Users of digital devices are now faced with "the self-regulatory cost of bringing your own boundaries" to attention.[16] For Crawford, "the left's project of liberation led us to dismantle inherited cultural jigs that once imposed a certain coherence (for better and worse) on individual lives."[17]

All three responses frame discrete user error as the problem: we users cannot be trusted with the tools we've been given. Like amateur musicians, we simply have poor training or undisciplined technique. In their efforts to humanize technology, digital minimalists frame the issue in terms of a public health crisis. End users are offered palliative care in the form of digital detox retreats, meditation apps, and quantified self-trackers for limiting screen time. I, for one, have often felt like I've needed these interventions. With the array of distractions in my pocket, ready to chase away the slightest hint of boredom with a swipe, it can feel like I've lost some mature faculty of self-possession. It's as if the pressure of the external world has been muted while my internal, imaginative life has become either impoverished or oversaturated.

On the one hand, this situation isn't new. Already in 1924 Siegfried Kracauer sensed that boredom was on the verge of extinction with the craze for portable wireless receivers and the coming of live radio programming: "Since many people feel compelled to broadcast, one finds oneself in a state of permanent receptivity, constantly pregnant with London, the Eiffel Tower, and Berlin. Who would want to resist the invitation of those dainty headphones? . . . Silent and lifeless, people sit side by side as if their souls were wandering far away." In giving ourselves over to the flow of other minds on the airwaves, wireless sets "allow oneself to be chased away." Without ever having to experience boredom—which for Kracauer "provides a kind of guarantee that one is, so to speak, still in control of one's own existence"—wireless users rarely had to confront themselves anymore.[18]

On the other hand, the challenges confronting technique in the twenty-first century are vastly different from those that have come before. Digital minimalism has emerged at a moment when the tech industry continues to invent new methods for transforming our habits and desires into profitable data. Rather than understanding the impact of these extractive technologies as the result of capitalism, digital minimalists interpret them as the absence of religion or self-control.[19] The self-help nature of digital minimalism is disconnected from a broader accounting of the tech industry's collective, social impacts. Instead, digital minimalists emphasize the sanctity of an idealized, individual human experience without ever thinking to address the stratification of those experiences. Laura Portwood-Stacer identifies this disconnect when she argues that people who refuse to use smartphones or who opt out of social media—a practice she refers to as "conspicuous non-consumption"—may end up "framing refusal as a performance of elitism, which may work against observers interpreting conscientious refusal as a persuasive and emulable practice of critique."[20] Unlike community technology, digital minimalism assumes at the outset that users have no agency and aren't interested in acquiring it. Community technology, in contrast, gives users the choice of how far down the technology stack they'd like to go, empowering them to learn more if they'd like to, or simply to rely on having access and being able to share it with their neighbors.

While some recent manuals for digitally opting out are better conceptualized than others—especially Jenny Odell's wonderful *How to Do Nothing*, in many ways a parody of the genre—digital minimalism is largely an argument against technology that forecloses collective action.[21] Digital minimalists often fashion themselves as twenty-first-century Luddites. But as Megan Ward argues, Luddism was originally a political movement and an economic argument, not a self-help motto or personal virtue: "Refusing to be ceaselessly on-call may increasingly become the province of the wealthy and powerful, a return to an era when recreational time was a luxury of the elites rather than a social right for the masses."[22] Unplugging or digitally detoxing doesn't absolve us of any ethical entanglements at a time when gig workers don't have the privilege to put their devices away, or when Black and Brown communities don't have a say in whether they will be surveilled by facial recognition systems or disproportionately imprisoned by pretrial sentencing algorithms.[23] Ruha Benjamin, who details these and other forms of digital inequity in *Race after Technology: Abolitionist Tools for the New Jim Code*, writes of

> Silicon Valley parents requiring nannies to sign "no-phone contracts" and opting to send their children to schools in which devices are banned or introduced slowly, in favor of "pencils, paper, blackboards, and craft materials." All the while I attend education conferences around the country in which vendors fill massive expo halls to sell educators the latest products couched in a concern that all students deserve access—yet the most privileged refuse it? . . . Social theorist Karl Marx might call tech personalization our era's opium of the masses and encourage us to "just say no," though he might also point out that not everyone is in an equal position to refuse, owing to existing forms of stratification.[24]

To cite the title of a documentary film, "offline is the new luxury."[25] However, read in the light of Benjamin's diagnosis of a world in which many people still don't have access to digital media while privileged people in the United States no longer want it, "offline is the new luxury" seems to be less a celebration of leisure and mindfulness than an accurate diagnosis of digital inequity. Digital minimalism advances an injunction to detox from tools that many don't have the luxury of saying no to, commodifying disconnection when many can't get online in the first place. Far from providing real relief or action against the incursion of the tech industry on our privacy and attention, those privileged enough to try digital detoxes can only ever experience a momentarily satisfying self-righteousness, as Alexis Shotwell describes in her book *Against Purity: Living Ethically in Compromised Times*.[26]

Efforts to make people more aware of the ways that digital distractions have gotten out of hand are of course commendable, but these individual techniques need to be joined by a collective accounting of their cumulative effects, social stratifications, and differential impacts. What we need is not seven habits for highly effective users; nor do we need digital abstinence or subtler notifications. Instead, we should ask which of these competing claims to minimalism—as a form of attention, mindfulness, and agency—are compatible with maximal connection, with maximal choice, and with maximal investment in communities and infrastructures. Our understanding of technique should encompass not only individual user and tool, but also what Mauss calls the "ensemble" of broader technical infrastructures supporting those devices as well as the values that communities hold in common.[27] We should understand technique today not only as a matter of individual choice but also in terms of its interwoven contexts: the factors influencing the way we do things.

Distraction felt like it was the central technique around which public discourse on tech revolved in the 2010s: the things we couldn't stop ourselves from doing, the fragmented attention we'll never put back together.[28] "I miss my pre-internet brain," reads a piece of installation art by Douglas Coupland that went viral in 2014. Now, automation is shaping up to be the central technique animating discourse on tech in the 2020s. It is shifting the terms we previously used to understand the relationship between users, tools, and techniques.

Automation is an idea with old roots, from automata and constrained writing to number games and computational thesauri.[29] In the eighteenth century, automation was already framed as a labor-saving device by an influential English parable: a boy working at the controls of a steam engine had the monotonous job of opening and closing the engine's valves: open the inlet valve to let steam into a cylinder, pushing a piston forward. Open the exhaust valve, releasing the steam to allow the piston to return to its starting point. Repeat. But all this boy wanted to do was go play. One day, finding a piece of string in his pocket, he was struck by an idea: he tied the inlet valve to one end of the piston and the exhaust valve to the other end. Now, when the piston reached the "top of its play," it opened one valve and closed the other, allowing it to return automatically at each end. The steam engine becomes a continuous circuit, and the line dividing labor from its automation is as thin as a piece of string.[30]

So far we're in familiar territory: someone lands on a creative technique for doing something more easily by knotting one mechanism to another and freeing up some time. Depending on the telling, the parable has been used to demonstrate the simplicity of cybernetics, the enduring creativity of humanity amid machines, and the ability of technology to make life easier. In some distant future, it might be the perfect parable for a utopian world without work, where it's no problem that steam engine operators everywhere are out of a job thanks to a postscarcity economy that provides for all.[31] But right now, it's less clear that people will have a say in when, how, and whether their income gets automated away from them.

Today, automation is less of a particular technique (a creative decision to modify a gadget) than it is the recording and capture of all technique. Take a familiar example. Not only have countless retail employees been replaced by now ubiquitous self-checkout terminals, but in the lead-up to that automation, employers subjected the on-the-job technique of those workers to exhaustive tracking and data capture, as Adam Greenfield describes: "Every checkout interaction at the Target chain, for example, is rated Green, Yellow or Red by an automated system, according to whether or not the clerk hit targets for speed and accuracy, and these ratings are used to determine employee compensation." Other corporations require employees to wear devices that track their location, water consumption, and interactions with customers.[32] Before these jobs were automated away, the most minute details about the way human workers did things was surveilled as if the companies took notes on how best to replace them.

From the parable of the boy and the steam engine and beyond, automation as a labor-saving device is an old idea. But AI's automation of cognitive and creative labor is certainly a new development, and one that is already having repercussions not only for work but also for the ways we communicate, socialize, and even read and write. All of this comes down to attempts to model the user and uses of a tool. Even before AI, platforms were able to tailor an application to individual users according to their past behavior. User experience design originally centered on relatively broad personas, or fictional users who a development and design team could imagine approaching their interfaces in particular ways. These personas helped designers build their tools around a delimited set of imagined use cases. Benjamin Bratton observes that eventually, these blunt personas were segmented into ever-smaller categories thanks to data on user behavior:

> Instead of abstracting a few typical personas out of the patterns seen through cumbersome, representative studies (we hope), and then designing offerings to suit those abstractions, . . . cloud platforms with access to each user's specific profile (previous search

history, purchases, geographic location, circle of friends) can hypersegment services and content. The ratio between number of users and possible personas is then trending toward a 1:1 ratio; your profile is your persona, and in principle, every data user gets his or her own version of any offering.[33]

In other cases, the process of abstraction moves in the opposite direction. Rather than abductively customizing technique to suit the known data on a specific user, AI systems inductively narrow an individual's range of possible actions based on the aggregate data of all recorded users. Click Reply in almost any email client and you will be offered a delimited set of autosuggestions, ranging from phrases that complete your thought ("I hope you . . . are doing well") to complete sentences that offer a sunny reply without having to enter more than a single keystroke ("Let me get back to you on that!"). These predictive text systems make assumptions about what you want to say based on massively aggregated models of user behavior. Given a language model generated from billions of emails, they generate a synthetic text based on the ways most correspondents have replied to similar messages. At best, these systems risk producing an echo chamber populated by stochastic parrots when writers are fed, select, and reproduce the same stock phrases and baseline tone, averaged out from the spectrum of possible responses.[34] At worst, they risk amplifying some of the most toxic and hateful sentiments that are all too often found on the internet, carelessly reproducing them in autosuggestions.[35]

On the surface, automated systems like these are designed to spare us precious time. No one actually likes replying to emails, so why not streamline the process with one-click options? The protagonist of Gish Jen's novel *The Resisters* (2020) follows this train of thought. Looking back from the perspective of a dystopian surveillance state, he asks how we got here:

> Like others, I had allowed Aunt Nettie [the novel's Big Brother] to keep my calendar back in the days when, as the young head of an English as a second language program, I had immigrants to

teach and obligations to juggle. This was some time ago, now—before Ship'EmBack. But back then, I had also allowed Aunt Nettie to email people on my behalf, checking the "mimic your voice" option and marveling at just how perfectly she could replicate my tics of phrasing. . . . As for why I did these things—I generally did them, I see now, because I appreciated some associated convenience.

We've all read dystopian warnings of Big Brother taking control. In this speculative near future, Jen wonders what if it's us who do it to ourselves, with tools we willingly invite into our lives?

And as for the resulting reality, was it not disconcertingly like the sea level rise and heat and wind we knew, long ago, would come with climate change but have since come to call normal? No one would have willfully chosen the stranding of whole office parks and schools and neighborhoods by the flooding we saw now. No one would have willfully chosen the generating of the places we called marooned places, just as no one would have chosen the extinction of frogs and of polar bears, or the decimation of our pine forests by the explosion in bark beetles. And yet it was something we humans finally did choose.[36]

Now, I get it. I do! It's impossible to keep up with notifications and tasks and the always-on expectations of remote work, and wouldn't it be great to automate all that away? To enjoy the leisure time that generative AI might free up for us? But the idea that there is a single most efficient—and thus correct—way to do things is an inherently technocratic view of the world. It risks not only narrowing our range of expressive possibilities but also encouraging conformity to some preconceived standard of performance. In both cases, automated systems make blunt assumptions about who you are and the way you do things. Your technique is predicted either on the basis of a walking zombie of misconceived metadata derived from you, or on the basis of calculations of every action performed by all users the system has witnessed in the past. The more our tools want to be used a certain way, the more they want to record the unique ways we use them, the farther we are from building systems based in democratic governance and local consensus. Instead, we have technocratically enforced decisions that are based on averages calculated at scale.

Digital minimalists design convoluted checks and balances that will prevent users from overusing their devices. The tech industry designs systems that either prescribe or automate the way we do things in the name of efficiency. In both cases, the real conversations about the stewardship of technology happen among Silicon Valley engineers, behind closed doors. We users remain distracted and in the dark, as if technique is out of our hands. But do we have to accept this? What if we don't want to enforce digital abstinence or count on the expertise of engineers? The answer is complicated, and it hinges on whether we understand ourselves—distracted users of information technologies—to possess technique.

Rarely do we consider the mundane activities of making tea, dusting a room, or raking leaves ever to achieve the qualities of technique.[37] We're even less likely to think of an activity like scrolling through our social media feeds as an example of technique. It feels closer to what William James describes as "the effortless custody of automatism:" the habits we rely on to go about our daily lives.[38] We casually use the word *technique* to describe the expert performance of cellists, surgeons, point guards, and other artisans. Scholarship on skilled artisans—especially by anthropologists and historians of science—has shown again and again that new epistemologies emerge from expert practice. These studies of sculptors, alchemists, and machinists reveal how technique is a mode of knowing things.[39] But what of the nonspecialist, the layperson? Can users of a smartphone, laptop, search engine—any digital platform—be skilled in the same sense?

The conventional response is no. Laypeople are "users" in

the sense of being addicted to some potent drug. Ellen Rose explains this common view of the user: "*using* is disassociated, by definition, from *knowing*, since the former term itself connotes a distinctly parasitic relationship with technology in which the user taps into and exploits a technology that has been created by more knowledgeable individuals, while contributing nothing in return."[40] Unlike artisans, users use things uniformly, which means that in this view, a smartphone can only be used too little or too much. It can't really be used differently from anyone else.

What makes this view troubling is not only that it paints users as comprising a class of subjects dependent on the vanishingly few people Delany describes as "know[ing] the incredibly complex spells and incantations needed to get the stuff to work."[41] Also troubling is that this portrait of a subservient user effectively spares no one. Given the ubiquity of computing interfaces in the modern world, everyone is a user. For Rose, it has become more and more difficult to draw a distinction between skilled technicians and amateur users: "in the early 1980s . . . one could choose to become a computer user, just as one's personal values, talents, or inclinations might lead one to become a gourmet cook, a downhill skier, a marathon runner, a poet." While this relatively small group of early computer users possessed specialized knowledge, Rose argues (back in 2003) that "today, becoming a computer user seems to be less a matter of choice than something that society requires of us."[42]

If Rose already sees everyone becoming a user at the turn of the twenty-first century, then the line between technician and user has only dissolved further thanks to the increasing scale and complexity of modern technical systems. Even software engineers at a large tech company are likely to specialize only in a small subset of the broader systems in which their code is just a part.[43] Similarly, the more convoluted AI becomes as it is fed an insatiable diet of image, music, and text, the less engineers are able to interpret or even explain the results produced by the systems they have created. For all the seemingly impressive achievements of AI, it still has a major interpretability problem. According to recent research from MIT's Lincoln Laboratory, "the machine learning process happens in a black box, so model developers often can't explain why or how a system came to a certain decision." This study shows that formal specifications—the current leading method of translating mathematical models into natural language descriptions of a system's outputs—are seldom accurate.[44]

From the vantage point of AI, there is a trade-off for users: not knowing brings ease. The entire point is to automate little inconveniences so that we're free to do more.[45] According to the AI imaginary, things are about to get a lot easier for everyone. As individuals, refrigerators will anticipate our nutritional needs and order our groceries; emails will reply for us so well that our inboxes might as well be talking to one another. But from a collective, distributed vantage point, the opposite feels true. The more groceries our refrigerator orders for delivery, the more it generates plastic waste and encourages precarious gig work. The more we rely on synthetic text to write emails for us, the more carbon outputs and precious water inputs are required to train the AI models that are just trying to sell us something (or sell us). So much effort is required to do almost anything conscientiously in the twenty-first century. Conscientious technique—ways of doing things that are responsive to the interconnected world we live in—requires an immense amount of conscious work, but the tools we use today are designed for unconscious, automated ease rather than deliberation over the relative merits of their use and their impacts.

So far, it seems like the deck is stacked against users of information technologies hoping to possess technique in any conventional sense. Yet even though the unconscious techniques of the user rarely involve deliberation, description, or decision, I cannot shake the idea that they are—like the use of a marlinspike to loosen a knot—brilliant spells and incantations in their own right. Consider the forms of procedural knowledge that users acquire when picking up a black box and figuring out how it works. Every time we swipe, pinch, double tap, long press, or hold to pause, we draw on a deep reserve of experiences that have culminated in a sophisticated repertoire of gestures for manipulating media of all kinds. We learn to use our devices in countless ways: the errant graze of a fingertip reveals what a two-finger swipe can do. A half-remembered, futuristic gadget from a science fiction film guides our intuition at an unfamiliar user interface.[46] The lag on a smartphone past its prime leads us to impatiently pull down to refresh a feed more times than is necessary. These unconscious techniques tend to change over time; the way you message a friend next year, for example, will almost certainly differ from the way you did so during the last. Like unspoken arguments over the ways tools might be used, techniques interact and evolve in a debate across individual forms of creativity and cultural tradition.[47] There are concepts embedded in the way we decide to use our tools, tacitly debated with others over time.

Philosopher Michael Polanyi refers to such concepts as "the tacit dimension," or the ways that "we know more than we can tell."[48] These ideas rarely look the same for any two people.[49] In our own ways, however, we self-forgetful users of distracting devices are technicians who, like the *Pequod*'s carpenter, possess unique forms of tacit knowledge as our neurons fire through the muscles of our fingers. We are like athletes prized for their ability to internalize a set of rules and "go unconscious" or "go automatic" when performing in a high-pressure situation. Redeeming what users know through embodied intelligence is vital during an age when our tools will become progressively more unknowable, even to experts and engineers.

An increasingly influential argument, in fact, holds that discussions of AI's interpretability problem ignore how users understand the emergent properties of AI systems in practice: "In context, inscrutability is not a result of technical complexity but rather of power dynamics in the choice of how to use those tools." Joshua Kroll calls this category error the fallacy of inscrutability.[50] The tech industry benefits from being able to obfuscate the details of AI systems behind a scrim of oracular, black-boxed insight that no one can possibly understand, let alone critique. But users have ways of understanding what a system is doing to them. Their intuition may actually be their greatest strength with AI. (What an inversion this would be! Engineers, unable to interpret the workings of their own systems, turning to the tacit ways that users know things!) Even though our seemingly inscrutable tools distract, record, and predict us, we improvise in order to understand and use them, bending them to suit our desires.

We may all be users, but that doesn't mean we are disempowered. As if anticipating the evolution of all people into users, philosopher Michel de Certeau proposes (as early as 1980) substituting the word *user* in place of the phrase *common man,* placing the everyday practices of users in a privileged position between producers and consumers in a capitalist society. Although users are "commonly assumed to be passive and guided by established rules," de Certeau describes everyday "ways of using the products imposed by a dominant economic order" as a form of "poaching," a tactical manipulation of the predefined circumstances in which we live.[51] Similarly, feminist historians of technology have shown how embodied know-how gives rise to ideas and arguments about technology. For scholars like Ruth Schwartz Cowan, Ruth Oldenziel, and Judy Wajcman, emphasizing users in the history of technology returns agency and visibility to the role of women as innovators. Until their scholarship, historians of technology studied invention and innovation, which led the field to be "dominated by stories about men and their machines," in the words of Nelly Oudshoorn and Trevor Pinch. "Feminist historians suggested that focusing on users and use rather than on engineers and design would enable historians to go beyond histories of men inventing and mastering technology."[52]

Embracing user technique as a flexible form of poaching or tactics will become all the more important as the smartphone—the totem for how we have lived with technology for almost two decades now—reaches an inflection point. In relatively short order, the smartphone became a needle's eye through which our entire world passed, largely by engulfing countless tools (address book, camera, newspaper, map, music player) into a single gadget. That gadget's days may be numbered. From a design perspective, the smartphone's mannerist phase is its most minimal and unadorned: a seamless black slab of edge-to-edge screen with no distinguishing features. Many futurists predict that the next twenty years will see a movement in the opposite direction through an explosion of new, unique devices "branching back out" from the smartphone: biometric rings, AI lapel pins, smartwatches, augmented reality glasses, even Bluetooth-enabled yoga pants.[53] Voice interfaces powered by conversational AI are making the prospect of a screen-free world seem plausible.

Before the release of this new post-smartphone tool kit, we could be driven away from our smartphones for other reasons. After the demise of Twitter as it morphed into X, social media feels entirely broken—and social media is always the killer app that keeps people coming back to their phones. Even the smartphone itself feels fragile. By this I don't just mean cracked screens or irreplaceable components fused to the case, making repair impossible. The fragility I'm talking about is the product of an ongoing cascade in the 2020s of supply chain shocks, climate crises, faltering energy grids, semiconductor chip races, and wars. It seems fair to wonder whether the global coordination of labor, resources, and infrastructures that made the smartphone possible will ever align again.

We've done a lot to accommodate the presence of the smartphone. We have even reconfigured our lives to suit it: I preemptively stash various chargers around my house, another at my office, and one in my car, all to avoid the dreaded low-battery icon. (It tugs at my awareness in the same way my car's gas gauge does when the needle sinks below ¼.) Without that phone, it's not only that I'm worried I'll miss a text or have to live with the urge to glance at the news. It's that without it, I won't be able to check in at a doctor's appointment or log in to my university account or deposit a check or show my ticket at a concert. Whatever new tools we end up incorporating into our lives will have limitations and frustrations of their own. The smartphone is only an interstitial gadget, like all the others, and its future is uncertain. Some will detox and decide to leave their phones behind; others will no longer be able to access or afford one; still others will buy ten gadgets rather than one. If the smartphone fades into obsolescence, perhaps we'll adopt an entirely new ensemble of techniques to suit these emerging devices sooner than we can imagine. As we have seen, the way we do things often exceeds the frame of the tools we use to do them. The embodied intelligence we so rapidly acquired for the smartphone will find new ways of taking root among other tools as we integrate them into our lives. Whatever tools come along, though, the challenge will be to find ways for the brilliance of the unconscious user to inform conscientious technique, and for the user experience of the individual to contribute to collective action.

Today is a work day at a park in the Kensington neighborhood of Philadelphia. It's spring, which means that members of the community garden are clearing their plots, transplanting tomato and herb seedlings, and mulching the paths. Others have gathered for a cleanup around the playground, which fills with the sound of kids at a newly installed jungle gym. They use sticks to drum resonant tones by striking multicolored metal cylinders. Volunteers with PCW have gathered at the park's northwest corner to check on some mesh nodes that seem slow. A couple of people are new to the group and ask how they can pitch in. Standing around a chicken coop, we introduce ourselves and talk through the operation of an adjacent, solar-powered mesh node that offers internet access to anyone sitting in the garden. It's still a prototype—a twenty-five-watt solar panel, recycled battery, and access point contained in a compact case—but we hope to make more of them soon, shoring up the mesh's ability to adapt around any signal loss or power failure.[54]

Like *pocket wireless* and its many competing descriptions a century earlier—aerophone, radiophone, Marconigram—*mesh* is currently a kind of floating signifier, the shape of which is being worked out by communities of practice like PCW. Others are actively working to take the word *mesh*—as an idea central to the work of community technology and its enactment of conscientious technique—and co-opt it to mean something else. Amazon sells a line of $400 mesh routers that propagate Wi-Fi signals within the closed walls of large, single-family homes. Other Amazon smart devices (doorbells, security cameras, Alexas) will share their connection to the internet with neighbors whose signal goes down—so long as they also own proprietary Amazon smart devices.[55] Facebook, Google, and SpaceX fly a mesh of balloons, drones, blimps, and satellites over the world's poorest areas to transmit the internet down from above, without ever having to set foot on the ground.[56] One startup even purports to offer a community mesh network, but in actuality, it sells $500 routers that mine cryptocurrency in exchange for attracting network users.[57]

Here at the park, we know that companies are out there beaming the internet down from space, seeking in exchange a steady flow of subscription fees, user data, crypto, and plaudits. But all of that seems far from mind. It's a beautiful spring day, and we have come to mend slow nodes. One person pulls back some soil with the heel of his shoe, showing the place where an Ethernet cable is sunk into the ground. We follow its path beneath several garden beds to the place where it exits on the other side of the plot. That cable connects a router to a directional antenna mounted atop a restaurant; it allows anyone dining there to connect, but more importantly, it serves as a relay for new mesh nodes on the adjacent block, growing the network to the west. There, we hope to recruit new volunteers and antenna hosts.

We inspect the restaurant's access point, which has been down for a couple of days now. It looks like it was just unplugged, probably by someone who needed an outlet for the electric leaf blower lying beside one of the garden beds. We plug it back in and walk to the next downed access point, where we find that it's waterlogged. We replace it with a fresh antenna and set the old one out to dry, resolving to come back in a couple of days to see if it still works.

A 1923 cartoon captioned, "In the year 2023 when all our work is done by electricity." Created by H. T. Webster, *New York World*.

EPILOGUE

Reclaiming Technique

AS THIS BOOK GOES TO PRESS, A CONFLUENCE OF FORCES called AI is quickly becoming a fact of life. The meaning of the term currently ranges from applied statistics (machine learning for medical diagnoses) to sentient robots (HAL 9000 turning on its human user).[1] This polysemy is in part what makes it so challenging to keep up with the pace of things. Depending on the framing, AI promises to do away with the annoying tasks we'd rather not do ourselves or with the labor that our employers would rather not pay us for. The advent of generative AI means that a mere two-sentence prompt can replace countless lines of computer code, generate a cinematic film recreating last night's dream, or produce a plausible interpretation of a literary text. In a piece contemplating "the end of the human writer," Samanth Subramanian says that "the stakes feel tremendous, dwarfing any previous wave of automation. Written expression changed us as a civilization; we recognize that so well that we use the invention of writing to demarcate the past into prehistory and history. The erosion of writing promises to be equally momentous."[2] It's difficult to see the things that AI already makes plausible and not question the world the next generation will need to be prepared for.[3]

Even though the underlying technologies have been in place for over a decade and trenchant scholarly accounts of their history, biases, and conceptual underpinnings are well established, AI's worldwide rollout happened very, very quickly.[4] With the public release of ChatGPT in November 2022—the platform that stoked the AI imaginary—just six months later the Writers Guild of America went on strike, in part to ensure that AI didn't replace human screenwriters.[5] Tech stocks rode a wave of AI hype to record highs. Seemingly everyone under the sun was using AI in some facet of their daily lives. I talked to friends and neighbors who confided in passing that they were using it in their work as doctors, store owners, lawyers, and teachers, relying on synthetic text to produce patient notes, product descriptions, legal briefs, and lesson plans. They all followed these admissions with a smirk and a shrug, as if to say: I know, it's terrible, but what are you going to do?

Beneath the breakneck rate of its adoption, AI's technical development began outpacing long-held benchmarks for computational change. Recent research shows that "LLMs are improving several times as fast as Moore's Law, the engine of the digital age, which holds that the number of transistors on a computer chip doubles every two years or so."[6] By the time this sentence goes from pixel to print, the situation described here will almost certainly seem quaint in light of the latest AI developments. Maybe when this book is ingested into the training data for the next large language model—returning from print to pixel—anyone will be able to create a GPT from its text and ask my book for thoughts on those latest developments.[7] For now, it's a safe bet that a lot of people are going to lose their jobs—people whom Gish Jen speculatively calls "the unretrainables"—and that some weird cultural shifts are on the way.[8] I've heard leading academic researchers sincerely suggest that, within a decade, AI may render meaningless some of our most vital interpersonal exchanges—not only trust that my interlocutor is human, but introduction, persuasion, and apology, or anything that gains its social currency through the cognitive labor required to perform it.[9]

I truly cannot decide how to feel about any of this. Some days I console myself with the thought that even Marcel Mauss—as we saw him amid the horrors of Vichy France—held out a utopian hope that great interdisciplinary teams of thinkers and technicians could coordinate the organization of complex technical systems for good: "the power of the instrument is the power of the mind and its use implies ethics and intelligence."[10] I rest securely in the knowledge that time and again, users have ultimately driven the development of new tools in the directions most suitable to them, and that the affordances of those new tools are always the product of cultural imaginaries with deep roots. Like pocket wireless over a century ago and countless media technologies before it, the emergence of AI is animated by visions of human augmentation. It feels like AI fell from the sky, but the dreams that it's made of have been passed down for generations. How many times have emerging media led to thoughts of transcending death? Ever since Félix Nadar's experiments of the 1850s, photography has been marked by the understanding that it captures and preserves individuals long after their passing.[11] RCA's famous logo, His Master's Voice, depicts a dog looking into the horn of a gramophone playing a recording of his owner speaking from beyond the grave.[12] We learned that early radio users conducted séances in the hopes of reaching the departed over the airwaves. Now, AI services offer deepfakes of loved ones we've lost, exquisite corpses generated using the text, photos, and videos they left behind.[13] All of this has been imagined before, and it's magic—not in Arthur C. Clarke's sense (complex technologies cannot be understood), but rather in Mauss's generous reading of magic as a shrewd negotiation between belief and understanding that breaks the link between means and ends. We don't really believe we're bringing someone back from the dead through the media's special effects, but the stagecraft of the attempt itself means something to us.

But even if mechanically reproducible sight and sound was a two-hundred-year rehearsal for the coming of AI, something feels different this time. The technology's creators are remorseful; visions of human augmentation have been supplanted by fears of human obsolescence. The cultural imperative driving the development of pocket wireless in the 1910s—a form of connection eagerly anticipated by the public before its technical possibility—is now a kind of economic imperative with AI.[14] "This is potentially the next trillion-dollar company"—a common refrain in press coverage—seems to be the prime mover. AI is not going anywhere, and it will shape culture and society in ways that are difficult to prepare for.

And so on other days, I feel like a battle was lost before anyone realized it had begun. Whether or not we users want these systems, we shepherd them into existence with each keystroke and every swipe. Technique is the fossil fuel of generative AI. Platforms once extracted and sold metadata about us—phone numbers, mailing addresses, ads we clicked, products we purchased. Now they extract the distinctive specificities of the way we do things in order to automate and reproduce the lilt in a speaker's voice, the facture of a painting, the style of this sentence. All are captured and amalgamated into one massive hyperobject calculated through trillions of parameters.[15] Tech leaders both pro and con seem to believe that AI is simply a law of nature that cannot be denied. A prominent venture capitalist, in the "Techno-optimist Manifesto"—a breathless defense of AI as a universal human good, an "engine of perpetual material creation, growth, and abundance"—mistakenly conflates the aims of tech with the intimacies of technique. He even claims the mantle of the ancient Greek term, writing that "*techne* has always been the main source of growth, and perhaps the only cause of growth, as technology made both population growth and natural resource utilization possible."[16] This manifesto collapses obvious distinctions between tools, our techniques for using them, the sociotechnical infrastructures in which they are embedded, and a specific form of unequal power relations that wields those infrastructures to extract capital.[17]

The fleeting, ephemeral nature of *tékhnē*—the craft, skill, and habit of knowing how to do something—could never be neatly pinned to a board like a lifeless specimen on display. But I remain troubled by the countless ways that I, and essentially anyone who has used the internet over the past decade, have unwittingly contributed to these models and the synthetic content they generate. No longer is my technique simply what makes me uniquely me. Now my technique could ultimately be the thing that makes me replaceable; worse yet, it could be just one lonely datum point enabling a vast model to render countless other people replaceable.

Perhaps we can imagine a future in which everything hinges on how singularly we can do things, and for how long. Sooner than is comfortable, it is possible that we will be forced to devise techniques that prove our humanity when the media we rely on for communication and understanding are flooded with synthetic content. Matthew Kirschenbaum has named this situation the textpocalypse: "a tsunami of text swept into a self-perpetuating cataract of content that makes it functionally impossible to reliably communicate in any digital setting."[18] What if, Kirschenbaum asks, we arrive at a place "where machine-written language becomes the norm and human-written prose the exception"? There are signs that we may in fact already be well on our way there.[19]

Just four months after the release of ChatGPT, user experience designer Maggie Appleton, in a widely circulated essay, recommends that we should "develop creative language quirks, dialects, memes, and jargon" for an internet polluted by generative content. Drawing on Saussure, she writes,

> We have designed a system that automates a standardised way of writing. We have codified *la langue* at a specific point in time. What we have left to play with is *la parole*. No language model will be able to keep up with the pace of weird internet lingo and memes. I expect we'll lean into this. Using neologisms, jargon, euphemistic emoji, unusual phrases, ingroup dialects, and memes-of-the-moment will help signal your humanity.[20]

Through this brilliant suggestion, we can see one possible future: a world in which users must continue to generate novelty in order to outsmart the AI systems seeking to learn from that novelty. The creation of new jargon may prove you're human for a while, but the half-life of those trends would always have to be faster than the efficiency of the scrapers assimilating all the new content. What this signifies is the beginning of a high-stakes race between technique and tech: users devise novel techniques to stay one step ahead of tech as it eats the valuable new information users leave in their wake.

If our goal is to distinguish human from machine or to outsmart AI's ability to adequately replace cognitive labor, then we can keep running, generating a torrent of new cultural forms in the process. Maybe, every now and then, we can look over our shoulders and toss a data-poisoning tool that invisibly alters a piece of content in ways that will break any AI model that tries to ingest it.[21] But we can't slow down lest the beast catch up. So in another possible future, we go elsewhere. Let's say the internet follows its current path, devolving into an even worse place than it already is through a synthetic flood of bland, nonsensical, toxic, or simply false content. We then might turn to locally managed intranets run by community organizations, to vibrant slow-text experiments with pre–World Wide Web networking protocols of the 1990s like Gopher, to long range, low-power radios for off-grid messaging, or to handwritten letter correspondence.[22] Then, as we starve it of fresh human data, we goad tech into eating its own tail as it is forced to use synthetic data to train its models, provoking a phenomenon called model collapse: "a degenerative learning process where . . . the model becomes poisoned with its own projection of reality."[23]

Take your pick of these two speculative fictions: a future of novelty outracing its appropriation, or a future where refusal actively creates conditions that demand a better information economy. What both visions share is a portrayal of a world where technique is still in our hands. As Mauss notes, "the tool is nothing if it is not handled." If AI renders the internet useless, then users will find other ways to be seen, heard, and valued. They will refuse systems that extract technique in order to replace it, and they will pour their creative energies and trust into something different.

—April 2024

ACKNOWLEDGMENTS

I WROTE THIS BOOK OVER A PERIOD OF SIX YEARS, BUT the seeds of its ideas were planted much earlier. It lived many lives across academic contexts and communities, as did I through it. I'm grateful to the countless people and institutions who supported me and this project over that period.

Early book drafts were workshopped by the First Book Institute at the Center for American Literary Studies and the NEH Object Lessons Institute at Arizona State University. I am deeply appreciative of Sean X. Goudie and Priscilla Wald (at the FBI) and Ian Bogost and Christopher Schaberg (of Object Lessons) for encouraging the academics at these workshops to think of themselves as writers in addition to scholars. They pulled my cohort out of the weeds of our field-specific arguments and citation networks, insisting that we consider the overarching narrative structure of our books.

At the Columbia University Society of Fellows in the Humanities, I benefited from an embarrassment of resources and a community of brilliant colleagues spanning the humanities disciplines. Thanks to Vanessa Agard-Jones, Teresa Bejan, Maggie Cao, William Derringer, Brian Goldstone, David Gutkin, Hidetaka Hirota, Murad Idris, Ian McCready-Flora, Dan-el Padilla Peralta, Carmel Raz, and Rebecca Woods, and to Christopher Brown and Eileen Gillooly for their leadership and guidance. Dennis Tenen and Alex Gil, my collaborators on the Group for Experimental Methods in the Humanities, left an indelible mark on my thinking, and I continue to draw on their inspiration today.

At Pennsylvania State University, I benefited from a vibrant community of postdoctoral scholars, graduate students, and librarians assembled by Eric Hayot at the Center for Humanities and Information. I am grateful for his stewardship of this community and to the following individuals for their friendship and feedback during my time in State College: Jonathan Abel, Jeffrey Binder, Christian Haines, Steve Hathaway, Michelle Nancy Huang, Carrie Neal Jackson, Derek Lee, Jocelyn Rodal, Tracy Rutler, Josh Shepperd, and Shannon Telenko.

I've now found an academic home with the Center for Digital Humanities at Princeton. To my colleagues at the CDH—Nick Budak, Happy Buzaaba, Gissoo Doroudian, Sierra Eckert, Wouter Haverals, Rebecca Sutton Koeser, Kavita Kulkarni, Zoe LeBlanc, Emily McGinn, Rebecca Munson, Mary Naydan, Christine Roughan, Carrie Ruddick, Elizabeth Samios, Jean Shaver, Jeri Wieringa, and Mana Winters—I am grateful for your friendship, collaboration, creativity, and brilliance, and for the new things you teach me literally on a daily basis. And thanks for all the tacos. I am especially grateful for valuable conversations on this book with my CDH colleagues Ryan Heuser and Laure Thompson, and to Natalia Ermolaev and Meredith Martin for their deeply rewarding mentorship. From them, I have learned more than I ever set out to learn. Elsewhere at Princeton, thank you to Eduardo Cadava, Anne A. Cheng, Brigid Doherty, Bill Gleason, Thomas Y. Levin, and Nikolaus Wegmann for their guidance on the earliest versions of these ideas.

This book begins during an influenza pandemic and ends a century later, in a moment of hope, during Covid-19. To my friends and neighbors at Philly Community Wireless: thank you for giving me that hope. I have learned so much from all of you—about consensus, organizing, strategy, networks, and appropriate technology—and I am inspired by your vision. Thank you David Berman, Will Dean, Allan Gomez, Jonathan Latko, Heather Lewis-Weber, Jessa Lingel, Michael A. Major, Chris Mehretab, Stasia Monteiro, Sascha Meinrath, Leanne Przybylowski, Eugene Ryoo, Mark Steckel, Devren Washington, and Alex Wermer-Colan, as well as the passionate community of volunteers and community members who sustain the network.

ACKNOWLEDGMENTS

As the visionary director of the University of Minnesota Press, Douglas Armato publishes scholarship that has both defined my field and inspired me personally since I started in this profession. It is an honor to be included among his published authors, and I am grateful to him for being such a champion of my work. My thanks to Shelby Connelly, Eric Lundgren, Zenyse Miller, Jeffery Moen, Maggie Sattler, Mike Stoffel, Carla Valadez, and the entire production team for their careful attention to detail as they tended to all the moving parts of making this book. Thank you to cover designer Michel Vrana for capturing the spirit of this book in an image and to indexer Paula Durbin-Westby for breaking the book down into its constituent ideas by being such a thoughtful interpreter and correspondent. A special shout-out to my friend Judy Gilats, the book typesetter who designed these pages. Thank you, Judy, for opening up your process and teaching me about typography and book design along the way.

Many people read early versions of this book in full or in part over the years, offering invaluable feedback. Because of these people, writing it felt like a dialogue: James J. Brown, Richard Dienst, Bernhard Geoghegan, Lisa Gitelman, Jeffrey West Kirkwood, Matthew Kirschenbaum, Meredith Martin (again, who could be listed in multiple categories of these acknowledgments), Shannon Mattern, Mara Mills, J. D. Porter, Jillian Sayre, Jason Stopa, and William J. Turkel. Two anonymous peer reviewers and the Electronic Mediations series editors—N. Katherine Hayles, Peter Krapp, Rita Raley, and Samuel Weber—helped me sharpen the book's arguments and see its place within the field.

Thanks especially to my old friend F.T. for the editorial guidance and the confidence; to Jonathan Fedors for a close friendship that found a way to grow during the pandemic through runs and socially distant sports broadcasts; and to Michael Joseph for workshopping countless ideas and inspiring countless others, and for sharing the endlessly fascinating things he finds in the world.

Finally, to my family: thank you for your love, support, and inspiration: Bob and Denise; Erin and Ben; Dan, Jen, and Harper; Alan and Carol; Marcella, Dan, Ella, and Luca; Anna, Josh, Matt, and John; Gail, Larry, and Alissa. To Sara and Asa: you two are my world and I love you both more than I can say. Sara J. Grossman: thank you for reading every word of every draft of every version of this book. As a writer, teacher, scholar, department leader, and tender of the natural world, you set a model that I seek to follow. You and I committed to this work and way of life when we were way too young to know what we were getting ourselves into. After countless jobs and moves, I thank you for continuing to choose it—and choose me—over and over again. Asa: you are a walking, talking force of nature. Witnessing you become yourself has been the single greatest privilege of my life.

NOTES

Preface: Technique in the Age of Tech

1. Hansen has written on the "missed encounter" between Foucault and media studies, and the possibility of theory at the scale of the individual that was papered over in his late work in favor of speculative concepts of governmentality and security. Mark B. N. Hansen, "Foucault and Media: A Missed Encounter?," *South Atlantic Quarterly* 111, no. 3 (2012): 497–528, https://doi.org/10.1215/00382876-1596254.
2. Ruth Schwartz Cowan, *More Work for Mother: The Ironies of Household Technology from the Open Hearth to the Microwave* (New York: Basic, 1985). Lucille Alice Suchman, *Human-Machine Reconfigurations: Plans and Situated Actions* (Cambridge: Cambridge University Press, 1987).
3. Tero Karppi, *Disconnect: Facebook's Affective Bonds* (Minneapolis: University of Minnesota Press, 2018); Trine Syvertsen, *Digital Detox: The Politics of Disconnecting* (Bingley, U.K.: Emerald, 2020). On refusal, see also Marika Cifor et al., "Feminist Data Manifest-No," 2019, https://www.manifestno.com. Brock's work bridges two typically distinct approaches in U.S. media studies with his media archaeological attention to technical detail and cultural studies of user communities: André Brock Jr., *Distributed Blackness: African American Cybercultures* (New York: NYU Press, 2020).
4. Reading Foucault and Giorgio Agamben's theories of *dispositif* (apparatus) as a series of "discourses, institutions, architectural forms, regulatory decisions, laws, administrative measures, scientific statements, philosophical, moral and philosophical propositions," Preciado argues that "within the Cold War regime, interior design, gadgets and multimedia techniques become 'pharmacopornographic apparatuses,' new governmental technologies of gender and sexual subjectivation." Paul Preciado, *Pornotopia: An Essay on Playboy's Architecture and Biopolitics* (Princeton: Princeton University Press, 2014), 85–86. For Sofia, "artifacts for containment and supply are not only readily interpreted as metaphorically feminine; they are also historically associated with women's traditional labors." Zoë Sofia, "Container Technologies," *Hypatia* 15, no. 2 (2000): 181–201 at 182.
5. Richard Sennett, *The Craftsman* (New Haven, Conn.: Yale University Press, 2009). Tim Ingold, *Being Alive: Essays on Movement, Knowledge and Description*

(New York: Routledge, 2011). Ian Hodder, *Entangled: An Archaeology of the Relationships between Humans and Things* (Malden, Mass.: Wiley-Blackwell, 2012).

6. Bernhard Siegert, "Cultural Techniques, or the End of the Intellectual Postwar Era in German Media Theory," *Theory, Culture, and Society* 30, no. 6 (2013): 48–65, at 62, https://doi.org/10.1177/0263276413488963. For an overview of *Kulturtechniken*, see Bernard Dionysius Geoghegan, "After Kittler: On the Cultural Techniques of Recent German Media Theory," *Theory, Culture, and Society* 30, no. 6 (2013): 66–82, https://doi.org/10.1177/0263276413488962. For an extended analysis of the concept of culture in cultural techniques, see Geoffrey Winthrop-Young, "The *Kultur* of Cultural Techniques: Conceptual Inertia and the Parasitic Materialities of Ontologization," *Cultural Politics* 10, no. 3 (2014): 376–88, https://doi.org/10.1215/17432197-2795741.

7. Joel B. Lande and Dennis Feeney, eds., *How Literatures Begin: A Global History* (Princeton, N.J.: Princeton University Press, 2021), 93–94. Edward Slingerland, *Effortless Action: Wu-wei as Conceptual Metaphor and Spiritual Ideal in Early China* (Oxford: Oxford University Press, 2003), https://doi.org/10.1093/0so/9780195138993.001.0001. Yuk Hui, *The Question Concerning Technology in China: An Essay in Cosmotechnics* (Cambridge, Mass.: MIT Press, 2016). Banú (sons of) Músà bin Shákir, *The Book of Ingenious Devices/Kitáb Al-Ḥiyal*, trans. Donald R. Hill (Amsterdam: Springer, 1979).

8. On the path from Mauss's theory of body techniques to the concept of *Kulturtechniken*, see Erhard Schüttpelz, "Body Techniques and the Nature of the Body: Re-reading Marcel Mauss," *Limbus: Australian Yearbook of German Literary and Cultural Studies* 3 (2010): 177–94; Winthrop-Young, "*Kultur* of Cultural Techniques." On the links between Leroi-Gourhan, Mauss, and Pitt-Rivers, see chapter 2 of this book.

9. On the history of translating the German concept of *die Technik* as *technique* and later as *technics*, see Eric Schatzberg, *Technology: Critical History of a Concept* (Chicago: University of Chicago Press, 2018), 147–51. For more on "the rather inhabitual *technics*," and why the awkwardness of the word is perhaps an advantage, see Samuel Weber, *Mass Mediauras* (Stanford, Calif.: Stanford University Press, 1996), 51–54.

10. "In the older handicraft-based world, Marx explained, a workman 'makes use of a tool,' that is, the tool had a metaphoric relation to the innate powers of the human subject. In the factory, Marx contended, the machine makes use of man by subjecting him to a relation of contiguity, of part to other parts, and of exchangeability." Jonathan Crary, *Techniques of the Observer: On Vision and Modernity in the Nineteenth Century* (Cambridge, Mass.: MIT Press, 1992), 131.

11. Cecile Malaspina and John Rogove, introduction to Gilbert Simondon, *On the Mode of Existence of Technical Objects*, trans. Malaspina and Rogove (Minneapolis, Minn.: Univocal, 2017), 15–21, at 15.

Chapter 1: Sleight of Hand

1. Articles covering MacFarlane's invention include "Radiotelephone Station in a Handbag," *Popular Mechanics* 31, no. 6 (1919): 807; "Pocket Wireless," *Independent*, July 1919, 11; Earl G. Hanson, "The Audio Frequency Radiophone," *Radio News*, November 1921, 381; "Call Up Wifey on the 'Stove-Pipe' Radio," *Electrical Experimenter*, June 1919, 115.
2. My sources in England tell me that the word *aerial* is still preferred there over *antenna*.
3. Schiffer, an archaeologist and radio historian, writes that pocket wireless sets like the MacFarlanes' would have largely been provisional prototypes in the 1910s, impossible to produce commercially. "These sets of ultimate portability obviously played poorly and were regarded as little more than novelties that advertised the cleverness of their makers." Michael B. Schiffer, *The Portable Radio in American Life* (Tucson: University of Arizona Press, 1991), 39. Schiffer argues that the pocket wireless is best understood as what he calls a cultural imperative: "a product fervently believed by a group—its constituency—to be desirable and inevitable, merely awaiting technological means for its realization." Michael Brian Schiffer, "Cultural Imperatives and Product Development: The Case of the Shirt-Pocket Radio," *Technology and Culture* 34, no. 1 (1993): 98–113, https://doi.org/10.2307/3106457.
4. A signal's wavelength is measured by the distance between the peaks of two successive waves.
5. "Radiotelephone Station in a Handbag."
6. Frank Chambers, another Philadelphia inventor, had demonstrated his pocket wireless to journalists eight years earlier. See H. D. Jones, "Pocket Wireless," *World Today* 20 (1911): 747. The Ondophone, a compact crystal detector, was made in Paris by Horace Hurm around 1914. See "Vest-Pocket Wireless Receiving Instrument," *Electrical Review and Western Electrician*, April 1914, 745. Pocket wireless sets were even depicted in silent films, where they were used as plot devices. See "Pocket Wireless Set in a Photoplay," *World's Advance* 30, no. 5 (1915): 618. Other portable wireless receivers are detailed in Schiffer, *Portable Radio*, 34–42. See also Erik Born, "Going Wireless in the Belle Epoque," *Continent* 7, no. 1 (2018), https://continentcontinent.cc/archives/issues/issue-7-1-2018/going-wireless-in-the-belle-epoque.
7. Hugo Gernsback, "Amateur Radio Restored," *Electrical Experimenter* 7,

no. 2 (1919), 131. The year before, *Electrical Experimenter*, one of the magazines interviewing MacFarlane, had been critical of the decision by the U.S. government to close the amateur radio airwaves.

8. The phrase "pattern of expectation" comes from Clarke's history of science fiction, a genre whose defining feature he takes to be depictions of the future. I. F. Clarke, *Patterns of Expectation, 1644–2001* (London: Jonathan Cape, 1979).

9. Quoted in John M. Barry, *The Great Influenza: The Story of the Deadliest Pandemic in History*, rev. ed. (London: Penguin, 2005), 478.

10. "It was common practice in 1918 for people to hang a piece of crepe on the door to mark a death in the house. 'There was crepe everywhere,' . . . recalled Anna Milani. 'We were children and we were excited to find out who died next and we were looking at the door, and there was another crepe and another door.' There was always another door. 'People were dying like flies,' Clifford Adams said. 'On Spring Garden Street, looked like every other house had crepe over the door. People was dead there.'" With nowhere to bury all the bodies, caskets were being stolen. Families had to leave bodies in a room "sometimes for days, the horror of it sinking in deeper each hour, people too sick to cook for themselves, too sick to move the corpse off the bed, lying alive on the same bed with the corpse." Barry, 223.

11. For more on social distancing measures adopted in U.S. cities during the influenza pandemic, see Richard J. Hatchett, Carter E. Mecher, and Marc Lipsitch, "Public Health Interventions and Epidemic Intensity during the 1918 Influenza Pandemic," *Proceedings of the National Academy of Sciences of the United States of America* 104, no. 18 (2007): 7582–87, https://doi.org/10.1073/pnas.0610941104; Nancy Tomes, "'Destroyer and Teacher': Managing the Masses during the 1918–1919 Influenza Pandemic," *Public Health Reports* 125 (2010): 48–62, https://www.jstor.org/stable/41435299.

12. For a discussion of "the symbolic import and cultural resonance of the etheric ocean as both a communications and social metaphor," see Jeffrey Sconce, *Haunted Media: Electronic Presence from Telegraphy to Television* (Durham, N.C.: Duke University Press, 2000), 59–72. For more on imaginary, supernatural, and occult media, see Stefan Andriopoulos, *Ghostly Apparitions: German Idealism, the Gothic Novel, and Optical Media* (Princeton, N.J.: Zone, 2013); Simone Natale, *Supernatural Entertainments: Victorian Spiritualism and the Rise of Modern Media Culture*, reprint ed. (University Park: Pennsylvania State University Press, 2017); Teresa De Lauretis, Andreas Huyssen, and Kathleen M. Woodward, *The Technological Imagination: Theories and Fictions* (Madison, Wis.: Coda, 1980). Kittler is emphatic on this connection: "A medium is a medium is a medium. As the sentence

says, there is no difference between occult and technological media." Friedrich A. Kittler, *Discourse Networks, 1800/1900*, trans. Michael Metteer and Chris Cullens (Stanford, Calif.: Stanford University Press, 1990), 229.
13. A phrase used in "Radio Is Blamed: Insanity Induced by Wireless Blamed for Deaths by Bible Teacher," *South Bend (Ind.) Tribune*, April 1923.
14. Kittler, *Optical Media*, 101. The passage from Kittler refers to the precinematic development of projected, moving images among spectacular magic lantern shows in the nineteenth century. "What Edison and the Lumières accomplished [with cinema] a century after Cagliosto and Schröpfer [with magic lantern shows] therefore fulfilled neither some timeless need nor some primal dream of humankind, which according to Zglinicki has supposedly been around from time immemorial; rather, it was a technical and thereby definitive answer to wishes that had been historically produced."
15. Guglielmo Marconi, "Syntonic Wireless Telegraphy," *Journal of the Society of the Arts*, May 1901, 516.
16. Sconce, *Haunted Media*, 60, 68.
17. Gernsback, "Amateur Radio Restored."
18. Sconce uses "electronic kinship" to describe the eerie experience of disembodied communication echoing through the air, as captured by storytellers in the first decades of the twentieth century. Sconce, *Haunted Media*, 62.
19. Ling Ma, *Severance* (New York: Farrar, Straus & Giroux, 2018), 31.
20. Mattern writes about the "Great Pause" during the pandemic, and the precarious, underpaid workers who made that pause possible: "the rush of activity that enables privileged retreat; the Othered precarity that ensures our security." Shannon Mattern. "How to Map Nothing," *Places Journal*, March 2021, https://doi.org/10.22269/210323.
21. Alvaro Sanchez, "Toward Digital Inclusion: Broadband Access in the Third Federal Reserve District," Federal Reserve Bank of Philadelphia, March 2020.
22. Adam Satariano and Davey Alba, "Burning Cell Towers, Out of Baseless Fear They Spread the Virus," *New York Times*, April 10 (updated April 11), 2020.
23. On the afterlives of nineteenth-century Luddism, see Matt Tierney, *Dismantlings: Words against Machines in the American Long Seventies* (Ithaca, N.Y.: Cornell University Press, 2019), 13. For the lessons the Luddites offer contemporary conversations on AI and labor, see Brian Merchant, *Blood in the Machine* (Boston: Little, Brown, 2023).
24. Susan Sontag, *Against Interpretation, and Other Essays* (New York: Picador, 2001), 172.
25. I first heard Mark Goble make this comparison in "End Time Forever: On Cinematic Slowness," Princeton University lecture, March 2012.
26. I am indebted to Heffernan for describing these substances in her talk:

Virginia Heffernan, "A New Aesthetic Called 'Aesthetic': Online Frescoes, Instagram Self-Portraiture and What Digital Humanities Can Be," Heyman Center for the Humanities, Columbia University, New York, September 2016.

27. Comments by Zuboff in an interview with John Naughton, "'The Goal Is to Automate Us': Welcome to the Age of Surveillance Capitalism," *Observer*, January 25, 2019. Their discussion is a profile of Shoshana Zuboff, *The Age of Surveillance Capitalism: The Fight for a Human Future at the New Frontier of Power* (New York: PublicAffairs, 2019).

28. Low theory is a concept introduced by Stuart Hall that has been taken up most prominently by Jack Halberstam and McKenzie Wark. Halberstam explains: "I take the term low theory from Hall's comment on [Antonio] Gramsci's effectiveness as a thinker. In response to [Louis] Althusser's suggestion that Gramsci's texts were 'insufficiently theorized,' Hall notes that Gramsci's abstract principles 'were quite explicitly designed to operate at the lower levels of historical concreteness.' Hall goes on to argue that Gramsci was 'not aiming higher and missing his political target;' instead, like Hall himself, he was aiming low in order to hit a broader target. Here we can think about low theory as a mode of accessibility, but we might also think about it as a kind of theoretical model that flies below the radar, that is assembled from eccentric texts and examples and that refuses to confirm the hierarchies of knowing that maintain the high in high theory." Jack Halberstam, *The Queer Art of Failure* (Durham, N.C.: Duke University Press, 2011), 16, https://doi.org/10.1215/9780822394358, citing Stuart Hall, *Stuart Hall: Critical Dialogues in Cultural Studies*, ed. Kuan-Hsing Chen and David Morley (London: Routledge, 1990). Wark has used the concept throughout her work, beginning with McKenzie Wark, *The Beach beneath the Street: The Everyday Life and Glorious Times of the Situationist International* (London: Verso, 2011).

29. Feminist historians of technology have provided exemplary models for understanding the tactics and creativity of technology in use. On the concept of the work process, see Cowan, *More Work for Mother*. On the mutual shaping of gender and technology, see Judy Wajcman, "Feminist Theories of Technology," *Cambridge Journal of Economics* 34, no. 1 (2010): 143–52. On user agency in general, see Ron Eglash et al., eds., *Appropriating Technology: Vernacular Science and Social Power* (Minneapolis: University of Minnesota Press, 2004); Nelly Oudshoorn and Trevor Pinch, eds., *How Users Matter: The Co-construction of Users and Technology* (Cambridge, Mass.: MIT Press, 2005).

30. Several recent studies centering on the users of particular technologies have inspired this study. Rankin wrote a study of the first personal

computer users in the 1960s and 1970s: Joy Lisi Rankin, *A People's History of Computing in the United States* (Cambridge, Mass.: Harvard University Press, 2018). Eubanks has explored the incursion of so-called AI into social services, exacerbating the harms of poverty: Virginia Eubanks, *Automating Inequality: How High-Tech Tools Profile, Police, and Punish the Poor* (New York: St. Martin's Press, 2018). See also McNeil's personal history of community and connection on late 1990s internet platforms: Joanne McNeil, *Lurking: How a Person Became a User* (New York: Farrar, Straus & Giroux, 2019).

31. As Guillory has detailed in his study on the origins of media as a concept, the difference between a poem and a painting in the nineteenth century was understood in terms of aesthetic choices surrounding the objects being represented. Only later were they were understood as differences in medium: "the distinction between poetry and painting looks very different when reconceived as the distinction between media (print and plastic art). The status of representation too is altered in relation to the category of medium, which directs our attention first to the material and formal qualities of different kinds of cultural expression and only second to the object of representation." John Guillory, "Genesis of the Media Concept," *Critical Inquiry* 36, no. 2 (2010): 321-62, at 346-47, https://doi.org/10.1086/648528. On the distinctions made by twentieth-century American art criticism and curation between "*mediums* (specialized subcategories of 'art,' including but not limited to those represented by MoMA's departments) and *media* (an imbroglio of cultural production of little aesthetic or political value)," see Anna Shechtman, "The Medium Concept," *Representations*, no. 150 (2020): 61-90.

32. On content, see Tess McNulty, "Content-Era Ethics," joint special issue of *Cultural Analytics/Post45*, April 2021.

33. Brian Merchant, *The One Device: The Secret History of the iPhone* (Boston: Little, Brown, 2017).

34. Paul Frosh, *The Poetics of Digital Media* (Cambridge: Polity, 2018), 8.

35. For writings on technique, habit, know-how, psychotechnics, the useful arts, and the *chaîne opératoire*, respectively, see Ernst Kapp, *Elements of a Philosophy of Technology: On the Evolutionary History of Culture*, ed. Jeffrey West Kirkwood and Leif Weatherby, trans. Lauren K. Wolfe (Minneapolis: University of Minnesota Press, 2018); Wendy Hui Kyong Chun, *Updating to Remain the Same: Habitual New Media* (Cambridge, Mass.: MIT Press, 2016); Jason Stanley, *Know How*, reprint ed. (Oxford: Oxford University Press, 2013); Jeremy Blatter, "Screening the Psychological Laboratory: Hugo Münsterberg, Psychotechnics, and the Cinema, 1892-1916," *Science in Context* 28, no. 1 (2015): 53-76, https://doi.org/10.1017/S0269889714000325; Eric Schatzberg, *Technology: Critical History of a Concept* (Chicago: University of

Chicago Press, 2018); André Leroi-Gourhan, *André Leroi-Gourhan on Technology, Evolution, and Social Life: A Selection of Texts and Writings from the 1930s to the 1970s*, ed. Nathan Schlanger (Bard Graduate Center, 2020).

36. For an intellectual history of the term, see Matthew Fuller, *Media Ecologies: Materialist Energies in Art and Technoculture* (Cambridge, Mass.: MIT Press, 2005), 2–5. Among the scholars reviving media ecology are Boczkowski and Mitchelstein, who argue that the digital "has become an environment, rather than a series of discrete technologies, that envelops and shapes virtually all major facets of everyday life." Eugenia Mitchelstein and Pablo J. Boczkowski, *The Digital Environment: How We Live, Learn, Work, and Play Now* (Cambridge, Mass.: MIT Press, 2021), ix.

37. Carolyn Marvin, *When Old Technologies Were New: Thinking about Communications in the Late Nineteenth Century* (Oxford: Oxford University Press, 1988), 8.

38. Donna Haraway, "A Cyborg Manifesto" (1985), in *Manifestly Haraway* (Minneapolis: University of Minnesota Press, 2016), 33.

39. "Zahlreiche Technikgeschichten kummern sich kaum um die Gebrauchsfrage oder setzen vielmehr stillschweigend voraus, dass sich die Verwendung einer Maschine ganz selbstverstandlich aus ihren technischen Eigenschaften ergibt. [Scores of technological histories pay little attention to the question of their use, or rather, they take it as a matter of course that the use of a machine arises organically out of its technical properties.]" Patrice Flichy, *Tele: Geschichte der modernen Kommunikation* (Frankfurt am Main: Campus und Éditions de la Fondation Maison des Sciences de l'Homme Paris, 1994), 13. For some thoughts on why a history of technology in use is needed, see David Edgerton, "From Innovation to Use: Ten Eclectic Theses on the Historiography of Technology," *History and Technology* 16, no. 2 (1999): 111–36, https://doi.org/10.1080/07341519908581961. Important for our purposes is Edgerton's eighth thesis: "Invention and innovation rarely lead to use, but use often leads to invention and innovation" (123). In a reimagination of the theory of affordances, a sociologist has updated the concept for the era of digital devices, apps, and automation: Jenny L. Davis, *How Artifacts Afford: The Power and Politics of Everyday Things* (Cambridge, Mass.: MIT Press, 2020), https://doi.org/10.7551/mitpress/11967.001.0001.

40. André Leroi-Gourhan, *Gesture and Speech*, trans. Anna Bostock Berger (Cambridge, Mass.: MIT Press, 1993), 114. A concept originally suggested by Leroi-Gourhan's mentor, Marcel Mauss, *chaîne opératoire* literally means "operational chain" or "sequence." As Schlanger puts it in an article on the archaeology of the ancient mind, the *chaîne opératoire* concerns "the becoming of material culture," the literal succession of material actions

undertaken to produce a tool. Nathan Schlanger, "Mindful Technology: Unleashing the Chaîne Opératoire for an Archaeology of Mind," in *The Ancient Mind: Elements of Cognitive Archaeology*, ed. Colin Renfrew and Ezra B. W. Zubrow (Cambridge: Cambridge University Press, 1994), 143–51. It's now an important concept throughout prehistoric archaeology for combining an analysis of the mundane, material aspects of everyday tasks—the "processes by which naturally occurring raw materials are selected, shaped, and transformed into usable products"—with more "ideational or symbolic considerations" of the prehistoric culture that created those tools (25). As a methodological framework, it allows for considerations of both manufacture and use, as well as "a theoretically informed commitment to understanding the nature and role of technical activities in past human societies" (27). Nathan Schlanger, "The Chaîne Opératoire," in *Archaeology: The Key Concepts*, ed. Colin Renfrew and Paul Bahn (London: Routledge, 2004), 25–31.

41. "Charles Keller, a pioneer in the anthropological study of cognition in practice, aptly calls [the *chaîne opératoire*] an 'umbrella plan,' an idiosyncratic constellation—peculiar to each practitioner—of stylistic, functional, procedural and economic considerations assembled specifically for the task at hand." Ingold, *Being Alive*, 35, citing Charles M. Keller, "Thought and Production: Insights of the Practitioner," in *Anthropological Perspectives on Technology*, ed. Michael B. Schiffer (Albuquerque: University of New Mexico Press, 2001), 33–45.

42. From remarks Ayrton delivered as chair of the Society of the Arts in London, following a lecture by Marconi, "Syntonic Wireless Telegraphy." Ayrton was an electrical engineer known for his mid-1890s experiments with electric arcs. For more on William and Hertha Ayrton's research on the electric arc, see Sungook Hong, *Wireless: From Marconi's Black-Box to the Audion* (Cambridge, Mass.: MIT Press, 2001), 161–63.

43. A 2003 full-page advertisement in *The Face* magazine, shared by David Beer at https://davidbeer.substack.com/p/the-bubble.

44. As claimed in a 1922 newspaper article on noted radio home brewer Leon W. Bishop: "Five or six years ago, he won a reputation as being more or less of a nut because he might often be seen walking about the streets with wires dangling from his hat and running down to a cane, while another wire trailed from one foot. Occasionally Bishop would hold out his cane, put one hand in his pocket, fumble with something—and announce that he was receiving a wireless message. Today almost anyone would know what he was doing, but five or six years ago the man who knew anything about wireless was an exception and unless Bishop took the time to let spectators 'listen in' to the dots and dashes, no one believed him." Jack

Binns, "Amateurs Race to Make Vest-Pocket Set," *New-York Tribune*, February 1922. The presence of information in the pocket was important to Bishop: Leon W. Bishop, *The Wireless Operator's Pocketbook of Information and Diagrams* (Lynn, Mass.: Bubier, 1911). Perhaps the development of the pocket wireless a few years later was simply the next logical step for Bishop in compact, portable information.

45. Rebecca Tuhus-Dubrow, *Personal Stereo* (New York: Bloomsbury Academic, 2017), 96. For a description of the dominant cultural practices that came to be associated with radio in its postamateur era as it entered the mainstream in the 1920s, see Susan J. Douglas, *Listening In: Radio and the American Imagination* (New York: Random House, 1999). "Listening in" was a technique that went through three distinct stages in the 1920s: first DXing ("tun[ing] in to as many faraway stations as possible"), then music listening, and finally story listening with the rise of network programs and serials (57–58). See also John Durham Peters, *Speaking into the Air: A History of the Idea of Communication* (Chicago: University of Chicago Press, 1999).

46. Clapperton Chakanetsa Mavhunga, ed., *What Do Science, Technology, and Innovation Mean from Africa?* (Cambridge, Mass.: MIT Press, 2017), 3-4.

47. Schatzberg, *Technology*, 20, 174. See also Leo Marx, "Technology: The Emergence of a Hazardous Concept," *Technology and Culture* 51, no. 3 (2010): 561–77.

48. "The tech giants have become so big that their moves can cause swings in the benchmark S&P 500 index, whose members are weighted by market capitalization. As of February 2, the top five, combined with Tesla Inc. and Nvidia Corp., accounted for more than 26% of the weighting of the S&P 500, according to S&P Global." Michael Wusthorn, "Five Big Tech Stocks Are Driving Markets. That Worries Some Investors," *Wall Street Journal*, December 21, 2021.

49. Mary F. E. Ebeling, *Afterlives of Data: Life and Debt under Capitalist Surveillance* (Berkeley: University of California Press, 2022). Sarah Lamdan, *Data Cartels: The Companies that Control and Monopolize Our Information* (Stanford, Calif.: Stanford University Press, 2022). Armelle Skatulski, "Algorithm in Command? Automated Decision-Making and the GDPR," *Autonomy*, September 23, 2020.

50. For Couldry and Mejias, this digital extraction is an expansion of the ways colonialism has appropriated human life for centuries: "Colonialism's sites of exploitation today include the very same West that historically imposed colonialism on the rest of the world. . . . What if the armory of colonialism is expanding? What if new ways of appropriating human life, and the freedoms on which it depends, are emerging?" Nick Couldry and Ulises A. Mejias, *The Costs of Connection: How Data Is Colonizing Human Life and*

Appropriating It for Capitalism (Stanford, Calif.: Stanford University Press, 2019), x. In a special issue of *Critical Inquiry,* Halpern et al. "complicate the widespread belief that data is valuable (just like minerals or energy resources)." They argue that "discourses of data extractionism affirm rather than critique data's status as sovereign and representative of the world." As a corrective, they offer the notion of surplus data: "data, collected to pinpoint and describe some event, action, or identity, is now tasked with *more;* joining a sea of other data points, it becomes a source for the constant derivation for 'better' insight, 'more efficient' systems. *Big* [as in 'big data'] does not capture this extension, much less its precipitate in the social. Always immanently providing *more,* data is in surplus. . . . This *more* is not necessarily more value." Orit Halpern, Patrick Jagoda, Jeffrey West Kirkwood, and Leif Weatherby, "Surplus Data: An Introduction," *Critical Inquiry* 48, no. 2 (2022): 197-210, at 200, 201, https://doi.org/10.1086/717320.

51. Schatzberg, *Technology,* 149.
52. Winthrop-Young writes that *Technik* could be translated into English as *technology, technique,* or *technics.* "This is not an ideal solution; in some instances my choice may well be the inferior one. However, since *Kulturtechniken* [cultural techniques] encompass drills, routines, skills, habituations, and techniques as well as tools, gadgets, artifacts, and technologies, *cultural techniques* remains the most appropriate term." Geoffrey Winthrop-Young, "Translator's Note" to Bernhard Siegert, *Cultural Techniques: Grids, Filters, Doors, and Other Articulations of the Real* (New York: Fordham University Press, 2015), xv.
53. "Almost all Continental languages have a cognate of technique that can be translated into English as 'technology.'" Schatzberg, *Technology,* 8.
54. Tim Ingold, *The Perception of the Environment: Essays on Livelihood, Dwelling, and Skill* (London: Routledge, 2000), 370.
55. Mavhunga, *What Do Science, Technology, and Innovation Mean from Africa?,* 3. "Are we certain that what we call 'Western' science, technology, and innovation is indeed Western in origin, ingredients, and rationality? After all, from the Greek occupation of Dynastic Egypt in 323 BCE to the European colonization of the nineteenth century and now to this era of 'big data,' there has been a long history of translation and mobility of African, Asian, and Islamic knowledge and practices via the medium of colonial occupation and domination. We should not be shocked that Europe's scientific revolution occurred after, not before, the colonization of the Americas and India" (2).
56. Roopika Risam, *New Digital Worlds: Postcolonial Digital Humanities in Theory, Praxis, and Pedagogy* (Evanston, Ill.: Northwestern University Press, 2018), 43. Murray and Hand refer to these practices of making do as instances of

"technological disobedience, a term coined by Cuban artist and designer Ernesto Oroza" (144). Padmini Ray Murray and Chris Hand, "Making Culture: Locating the Digital Humanities in India," *Visible Language* 49, no. 3 (2015): 140-55. See also AfriGadget: Solving Everyday Problems with African Ingenuity (blog), accessed January 18, 2023, http://www.afrigadget.com/.

57. "This hidden polysemy has contributed to a pernicious conflation of meanings that tends to reduce the whole of the industrial arts to invention, and invention to applied science. This conflation has profound ideological implications, helping sustain a mystifying, deterministic discourse that portrays techno-logical change as the inevitable fruit of scientific discovery." Eric Schatzberg, "Technik Comes to America: Changing Meanings of Technology before 1930," *Technology and Culture* 47, no. 3 (2006): 486-512, at 512, https://doi.org/10.1353/tech.2006.0201.

Chapter 2: *The Way We Do Things*

1. Quoted in Marcel Fournier, *Marcel Mauss: A Biography*, trans. Jane Marie Todd (Princeton, N.J.: Princeton University Press, 2015), 344. The gathering on June 23, 1941, inaugurated a new colloquium Meyerson had recently founded in Toulouse, Société d'Etudes Psychologiques. Meyerson built a community of refugee scholars there in the free zone while also running an underground newspaper, *Armée Secrète du Sud-Ouest* (Secret Southwest Army).
2. Michael Robert Marrus and Robert O. Paxton, *Vichy France and the Jews* (New York: Basic, 1981), 3. Donna F. Ryan, *The Holocaust and the Jews of Marseille: The Enforcement of Anti-Semitic Policies in Vichy France* (Bloomington: University of Illinois Press, 1996), 23. In 1940, "the Pétain regime enjoyed widespread popularity, for most French people believed that Germany had won the war and that Britain would collapse imminently, making accommodation to Germany the only logical course to follow." Ryan, 20.
3. Carole Fink, *Marc Bloch: A Life in History* (Cambridge: Cambridge University Press, 1989), 251-52. Mauss served as an interpreter during World War I. His cousin (and Emile Durkheim's son), André, was killed in battle. Fournier, *Marcel Mauss*, 178.
4. Fink, *Marc Bloch*, 261-63.
5. Bloch had to convince an ultraconservative dean of faculty at Montpellier, who was not only sympathetic to Vichy's anti-Semitic legislation but also held a grudge against Bloch over a negative book review. Fink, 264-65.
6. Mauss held appointments at the École Pratique des Hautes Études and the Collège de France. Fournier, *Marcel Mauss*, 333, 338.

7. Quoted in Fournier, 336.
8. Despite the pain in his hands, Mauss insisted on sewing the required yellow star into his own coat. Fournier, 346.
9. Quoted in Fournier, 346.
10. Simon Kitson, *Police and Politics in Marseille, 1936-1945* (Leiden: Brill, 2014), 67-68.
11. The Commissariat Général aux Questions Juives was led by Xavier Vallat. Fink, *Marc Bloch*, 275.
12. Burke describes the approach of this group, sometimes referred to as *la nouvelle histoire*, as marking the French historical revolution. Peter Burke, *The French Historical Revolution: The Annales School, 1929—2014*, 2nd ed. (Cambridge: Polity, 2015).
13. Braudel, acknowledged as the leader of the Annales school's second generation, published notes on the conference proceedings a few years later. Fernand Braudel, "Note dans le Journal de Psychologie (1948)," *Annales* 6, no. 2 (1951): 242. The proceedings themselves were published as Ignace Meyerson et al., "Le Travail et les Techniques," special issue of *Journal de Psychologie* 41 (1948).
14. Marc Bloch, *French Rural History: An Essay on Its Basic Characteristics*, trans. Janet Sondheimer (Berkeley: University of California Press, 1966), 50, 52.
15. Mauss coauthored this treatise on magic with a friend of his, archaeologist Henri Hubert. Marcel Mauss, *A General Theory of Magic*, 2nd ed. (London: Routledge, 2001), 175.
16. Mauss exempts Hubert, who was "by profession a technologist." Marcel Mauss, *Techniques, Technology, and Civilization*, ed. Nathan Schlanger (New York: Durkheim, 2014), 50. Originally published as "Divisions et Proportions des Divisions de la Sociologie," *Année Sociologique*, n.s., 1927:98-176.
17. Marc Bloch, "Problèmes d'histoire des techniques," *Annales* 4, no. 17 (1932): 482-86, https://doi.org/10.3406/ahess.1932.1340.
18. Lucien Febvre, "Réflexions sur l'histoire des techniques," *Annales* 4, no. 36 (1935): 531-35.
19. For more on the psychotechnics of Hugo Münsterberg, which involved psychological experiments using short films produced in collaboration with Paramount Pictures in 1916, see Jeremy Blatter, "Screening the Psychological Laboratory: Hugo Münsterberg, Psychotechnics, and the Cinema, 1892-1916," *Science in Context* 28, no. 1 (2015): 53-76, https://doi.org/10.1017/S0269889714000325.
20. "Humans at work are better understood through the history of work and of techniques: a material history that is at the same time a social, moral, and psychological history. [L'homme au travail se comprend mieux par l'histoire du travail et des techniques: histoire matérielle et en même temps

histoire sociale, morale et psychologique]" (10). Ignace Meyerson, "Le Travail: Une Conduite," *Journal de Psychologie Normale et Pathologique* 41 (1948): 7-16. Only Bertrand Gille and Maurice Daumas, peripheral to historiography in France, took up this charge until the 1980s, when social scientists and historians of technology began collaborations that would grow into science and technology studies. The seminal essay collection is Wiebe Bijker, Thomas P. Hughes, and Trevor Pinch, eds., *The Social Construction of Technological Systems: New Directions in the Sociology and History of Technology* (Cambridge, Mass.: MIT Press, 1987). For more, see François Jarrige and Raphaël Morera, "Technique et imaginaire," *Hypothèses* 9, no. 1 (2006): 163-74, https://doi.org/10.3917/hyp.051.0163.

21. Beginning with his coauthored book on magic: Mauss, *General Theory of Magic*.

22. Emile Durkheim, *Lettres à Marcel Mauss*, ed. Philippe Besnard and Marcel Fournier (Paris: Presses Universitaires de France, 1998), 329-32. Durkheim wrote to his nephew offering him a loan to keep the venture afloat, concluding, "Please tell me you understand the mistake you've made. That you haven't seen things clearly. That will at least give me some hope. [Je t'en prie, prends conscience de la faute que tu as commise. Dis-moi que tu n'as pas vu les choses sous leur aspect véritable. Cela m'apportera du moins quelque espoir]" (332).

23. Mauss, *Techniques*, 79.

24. Marc Bloch, *Memoirs of War, 1914-15*, trans. Carole Fink (Ithaca, N.Y.: Cornell University Press, 1980), 14, cited by Katherine Stirling, "Rereading Marc Bloch: The Life and Works of a Visionary Modernist," *History Compass* 5, no. 2 (2007): 525-38, https://doi.org/10.1111/j.1478-0542.2007.00409.x.

25. Bloch, *French Rural History*, xxx.

26. Mauss, *Techniques*, 82. Hendren, a design and disability scholar, echoes this understanding of the body as an instrument: "Our bodies are not just the sacks of flesh that hold our 'real' intellectual selves; they are not fixed entities but mind-bogglingly adaptive, responsive instruments." Hendren identifies the normative assumptions that are baked into the design of the built environment, assumptions that determine how bodies *should* interact with the world. Instead, she suggests that designers and planners advance a form of assistance "based on the body's stunning capacity for adaptation, rather than a rigid insistence on 'normalcy.'" Sara Hendren, *What Can a Body Do? How We Meet the Built World* (New York: Riverhead, 2020), 29.

27. Originally published as Marcel Mauss, "Techniques du Corps," *Journal de Psychologie* 32 (1935): 271-93. The essay was an address to the Société de Psychologie Française on May 17, 1934. For a detailed account of the theory of body techniques and the history of its reception, see Erhard Schüttpelz,

"Body Techniques and the Nature of the Body: Re-reading Marcel Mauss," *Limbus: Australian Yearbook of German Literary and Cultural Studies* 3 (2010): 177-94.
28. Mauss, *Techniques*, 83.
29. Quoted in François Vatin, "Mauss et la technologie," *Revue du Mauss* no 23, no. 1 (2004): 418-33, https://doi.org/10.3917/rdm.023.0418.
30. Quoted in Schlanger's introduction to Mauss, *Techniques*, 21.
31. Mauss is paraphrasing Maurice Halbwachs, another sociologist associated with the Annales school. Marcel Mauss, "Conceptions which Have Preceded the Notion of Matter" (1939), in *Techniques*.
32. Jonathan Crary, *Techniques of the Observer: On Vision and Modernity in the Nineteenth Century* (Cambridge, Mass.: MIT Press, 1992), 16-17. While Crary never directly defines technique in the book, he uses the word to refer to material devices like "3-D movies and holography" (127) as well as ways of seeing that are conditioned by devices and social theories alike: "techniques for imposing visual attentiveness, rationalizing sensation, and managing perception" (24).
33. In an essay detailing "technological modifications by people with disabilities," Mills and Sterne seek to expand listening techniques beyond modes derived from the experience of the "nondisabled, normate subject" (404). Mara Mills and Jonathan Sterne, "Aural Speed-Reading: Some Historical Bookmarks," *PMLA* 135, no. 2 (2020): 401-11, https://doi.org/10.1632/pmla.2020.135.2.401.
34. Chun argues that digital networks necessitate a reappraisal of nineteenth-century theories of "habit" as a form of "unreflective spontaneity" in which "repetition breeds skill" (9). Chun points out that "habit itself is changing: it is increasingly understood as addiction." In order to comprehend the presence of digital networks in our lives, Chun suggests that we "refram[e] habit as publicity, as the remnants—or even scars—of others that one shelters within the self" (171). Wendy Chun, *Updating to Remain the Same: Habitual New Media* (Cambridge, Mass.: MIT Press, 2016).
35. Bloch's lecture at the Toulouse conference was published as Marc Bloch, "Les Transformations des Techniques Comme Problème de Psychologie Collective," *Journal de Psychologie Normale et Pathologique* 41 (1948): 104-19.
36. Bloch, "Les Transformations des Techniques," 108-9.
37. Bloch's economic history comparing techniques of the sixth and seventeenth centuries brings to mind Marx's comment: "Relics of bygone instruments of labour possess the same importance for the investigation of extinct economic forms of society, as do fossil bones for the determination of extinct species of animals. It is not the articles made, but how they are made, and by what instruments, that enables us to distinguish different

economic epochs." Karl Marx, *Capital: Volume One* (1867), trans. Samuel Moore and Edward Aveling, Marx/Engels Internet Archive (Marxists.org), https://www.marxists.org/archive/marx/works/1867-c1/.

38. Bloch, "Problèmes d'histoire des techniques." Braudel, on this same book, writes: "The mistake of Lefebvre des Noëttes, still in many ways an admirable writer, was to reduce the history of technology to a simple minded materialism. It just will not do to say that the horse-collar, which replaced the yoke-harness from about the ninth century, thus increasing the traction-power of horses, 'progressively reduced man's slavery.'" Fernand Braudel, *The Structures of Everyday Life: Civilization and Capitalism, 15th-18th Century, Volume 1* (New York: Harper & Row, 1982), 334.

39. Bloch, "Problèmes d'histoire des techniques," 55.

40. Bernhard Siegert, *Cultural Techniques: Grids, Filters, Doors, and Other Articulations of the Real*, trans. Geoffrey Winthrop-Young (New York: Fordham University Press, 2015), 12. Cornelia Vismann, "Cultural Techniques and Sovereignty," *Theory, Culture, and Society* 30, no. 6 (2013): 83–93, at 84, https://doi.org/10.1177/0263276413496851.

41. Sybille Krämer and Horst Bredekamp, "Culture, Technology, Cultural Techniques—Moving beyond Text," *Theory, Culture, and Society* 30, no. 6 (2013): 20–29, at 21, https://doi.org/10.1177/0263276413496287.

42. Febvre's preface to Bloch, *French Rural History*, xix.

43. Mauss, *Techniques*, 149.

44. Mauss devoted several years of his famous course on ethnography at the École Pratique des Hautes Études, particularly 1935-36, to the subject of technology. André Leroi-Gourhan and Georges Haudricourt were students in this class, and both spoke fondly of it in interviews. Mauss's notes for these lectures, with the gaps filled in by a student of the course, Denise Paulme, would become Marcel Mauss, ed., *The Manual of Ethnography* (New York: Berghahn, 2009). For a comprehensive overview of Mauss's previous writings on technique, scattered throughout his oeuvre, see Vatin, "Mauss et la technologie." Many of these texts have been collected and translated in Mauss, *Techniques*.

45. An archaeologist published a monograph on Hubert: Laurent Olivier, *La memoire et le temps: L'ouvre transdisciplinaire d'Henri Hubert* (Paris: Demopolis, 2018).

46. "De nos jours, la technique la plus élémentaire, par exemple celle de l'alimentation (nous en savons quelque chose en ce moment), rentre dans ce grand engrenage des plans industriels [Nowadays, even the most elementary technique, such as that of food production (we know something of this at the moment), is falling into this great gear of industrial plans]." Marcel Mauss, "Les techniques et la technologie (1948)," ed. François Vatin, *Revue*

du Mauss no 23, no. 1 (2004): 434-50, at 438, https://doi.org/10.3917/rdm.023.0434.
47. It's significant that Mauss takes not only scholars as his point of departure but also technicians and popular historians as an inspiration for technology as a field of study. Figuier was a pharmacist from Montpellier and the founder of scientific journalism in France, who popularized scientific knowledge to the public through exhibitions and the illustrated series *Les Merveilles de la Science* (1867-91) and *Les Merveilles de l'Industrie* (1873-77). A father-and-son physicist team together published a history: Antoine César Becquerel and Edmond Becquerel, *Résumé de l'histoire de l'electricité et du magnétisme, et des applications de ces sciences à la chimie, aux sciences naturelles et aux arts* (Paris: Librairie de Firmin Didot frères, 1858). Edmond's son, Henri, would win the 1903 Nobel Prize, along with Marie and Pierre Curie for their discovery of radioactivity. François Vatin notes that Mauss's childhood memories of Durkheim temper our accepted understanding of the latter as someone entirely indifferent to the material, technical details of industrial labor. Vatin, "Mauss et la technologie."
48. W. S. F. Pickering, "How Compatible Were Durkheim and Mauss on Matters Relating to Religion? Some Introductory Remarks," *Religion* 42, no. 1 (2012): 5-19, https://doi.org/10.1080/0048721X.2011.637313.
49. Mauss, *Techniques*, 149. Mauss revised the version of this talk he mailed to the conference over the course of 1941-43, as revealed in further correspondence with Meyerson. An annotated version of the original French text appeared as Mauss, "Les techniques et la technologie (1948)." Cf. Mauss's definition of technique in his lecture notes: "traditional actions combined in order to produce a mechanical, physical, or chemical effect, these actions being recognised to have that effect." Mauss, *Manual of Ethnography*, 24.
50. I jotted this line down after seeing it in someone's conference talk slides attributed to Taggart. Neglecting to note down the source or the presenter's name led me to read every Taggart book I could get my hands on, but I had no luck finding the line. This led me to write Taggart a letter. Taggart responded: "Sad to say, I don't think I can help you to anything I may have said about technique. Which is to say I don't recall any such statement. At a guess, it may be in one of my essays, especially those on Zukofsky. My essays (most of them) can be found in *Songs of Degrees*. Not in that book, there is a long essay on Melville's *Pierre* in *Arizona Quarterly*. Also, I have a dim recollection that a talk I gave on George Oppen in NYU was published online by the sponsor (Dactyl Foundation). Best of luck with your book." Letter to author, December 21, 2019. We'll keep the line his for now.
51. Mauss, *General Theory of Magic*, 25.
52. Mauss, 25.

53. Arthur C. Clarke, *Profiles of the Future: An Inquiry into the Limits of the Possible* (New York: Harper & Row, 1962).
54. Mauss, *General Theory of Magic*, 75.
55. Nicolas Nova et al., *Curious Rituals: Gestural Interaction in the Digital Everyday* (Pasadena, Calif.: Near Future Laboratory, 2012).
56. Taina Bucher, *If . . . Then: Algorithmic Power and Politics* (Oxford: Oxford University Press, 2018), 109.
57. Schatzberg, *Technology*, 63.
58. Nathan Schlanger, "Techniques as Human Action: Two Perspectives," *Archaeological Review from Cambridge* 9, no. 1 (1990): 18–26.
59. In this model, the more ornamented the tools, the more complex the prehistorical culture that produced them. Pitt-Rivers draws on Darwin's concept of unconscious selection—whereby, for instance, a farmer selects traits that are desirable for his own crop growth (like rapid seed germination and simultaneous ripening), and in doing so unconsciously contributes to a much broader evolutionary divergence of domestic seed crops from their wild plant precursors—in order to describe a process he calls cultural evolution: "The uncertain fractures of flint, the various curves of the trees out of which they constructed their clubs, and the different forms of bones, would lead them imperceptibly towards the adoption of fresh tools. Occasionally some form would be hit upon, which in the hands of its employer would be found more convenient for use, and which, by giving the possessor of it some advantage over his neighbours, would commend itself to general adoption." Augustus Henry Lane-Fox Pitt-Rivers, *The Evolution of Culture and Other Essays* (Oxford: Clarendon, 1906), 96.
60. Helmut Müller-Sievers, *The Cylinder: Kinematics of the Nineteenth Century* (Berkeley: University of California Press, 2012).
61. Jeffrey West Kirkwood and Leif Weatherby, "Introduction: The Culture of Operations: Ernst Kapp's Philosophy of Technology," in *Elements of a Philosophy of Technology: On the Evolutionary History of Culture*, 3rd ed. (Minneapolis: University of Minnesota Press, 2018). Leander Scholz, "Der Weltgeist in Texas: Kultur und Technik bei Ernst Kapp," *Zeitschrift für Medien- und Kulturforschung* 1 (2013): 171–90.
62. Espinas was an important figure for the beginnings of sociology and "introduced the idea of collective conscience to sociology by assimilating society to a living organism" in his most important work, *Les Sociétés Animales* (Animal societies; 1877). Fournier, *Marcel Mauss*, 23. Like Mauss, Espinas plays with the connotations of the term *technique*, finding it to have a "rather restricted meaning in French; we speak about teaching *technique*, about such or such manufacturing *technique* and we thus refer to the operative processes, or in general, to the special parts of industrial arts . . . rather

than to the arts themselves." Espinas instead prefers the word *practice* to *technique*, hence his naming of *praxiology*: "The word practice probably contains a wider meaning; it can easily be taken as a concrete substantive (practice, practices), it is suitable for all collective manifestations of the will, whether spontaneous or well thought-out. It provides an excellent term to refer to the science of that order of facts as a whole: Praxiology." Alfred Espinas, "The Origins of Technology," in *The Roots of Praxiology: French Action Theory from Bourdeau and Epinas to Present Days*, ed. Victor Alexandre and Wojciech Gasparski, trans. Catherine Schnoor (New Brunswick, N.J.: Transaction, 2000), 45–91, at 47. This text is a translation of Alfred Espinas, *Les origines de la technologie* (Paris: Ancienne Librairie Germer Baillière, 1897), 8.

63. Fournier, *Marcel Mauss*, 24.
64. Espinas, *Les origines de la technologie*, 7. Today, praxiology lives on in the world of business ethics, management studies, and operations research. An ongoing book series published by the International Society of Business, Economics, and Ethics—Praxiology: The International Annual of Practical Philosophy and Methodology—draws its inspiration directly from the work of Espinas. The field of praxiology today prizes, above all, "efficacy— composed of effectiveness and efficiency—[as] the basic criterion of good work in all areas of human activity. The presence of this praxiological argument supplemented with ethicality, known as the triple E (effectiveness, efficiency, ethicality)." Wojciech W. Gasparski, ed., *Praxiological Essays: Texts and Contexts*, vol. 25 of Praxiology: The International Annual of Practical Philosophy and Methodology (London: Routledge, 2018), x.
65. Mauss, *Techniques*, 149–50. Espinas also breaks down technology as a field of study into three branches, but along different lines: "Technology comprises three sorts of problems, resulting from three different ways of considering techniques. . . . The three studies put together make up general Technology *[la Technologie générale]*." First, "an analytical description of the arts, such as they exist at a given time, in a given society, to determine their various categories." Second, "under what conditions and according to what laws, each body of rules comes into play." And finally, "studying the future of these very organs, focusing either on the birth, the peak and the decline of each of them in a given society, or on the evolution of all of mankind's series of techniques from the simplest to the most complex, through the alternatives of tradition and invention which seem to constitute its rhythm." Espinas, "Origins of Technology," 47–48; Espinas, *Les origines de la technologie*, 8–9.
66. Mauss, *Techniques*, 152–53.
67. Mauss, *Techniques*, 153.

68. Fink, *Marc Bloch*, 260.
69. Eugen Joseph Weber, *My France: Politics, Culture, Myth* (Cambridge, Mass.: Belknap, 1991), 244. *Les Cahiers politiques* has been digitized by the Bibliothèque nationale de France and is available online at https://gallica.bnf.fr/ark:/12148/cb34400547n/date.
70. Fournier, *Marcel Mauss*, 348.
71. Fifty years after Bloch's brief comments on the plow—its technical shape and social context—Gille wrote a similar piece on the wheelbarrow. Bertrand Gille, "Petites Questions Et Grands Problèmes: La Brouette," in *La Recherche en Histoire des Sciences* (Paris: Éditions du Seuil, 1983), 79-88. For a special issue commemorating the new English translation of Simondon's *On the Mode of Existence of Technical Objects*, see Conor Heaney, "The Disparity between Culture and Technics," *Culture, Theory, and Critique* 60, no. 3-4 (2019): 193-204, https://doi.org/10.1080/14735784.2019.1689626.
72. See especially Winthrop-Young, "*Kultur* of Cultural Techniques." But see also Siegert, *Cultural Techniques*; Vismann, "Cultural Techniques;" and Geoffrey Winthrop-Young, "Cultural Techniques: Preliminary Remarks," *Theory, Culture, and Society* 30, no. 6 (2013): 3-19, https://doi.org/10.1177/0263276413500828.
73. From a 1927 text on "The Divisions of Sociology" published in *Année Sociologique*. Translated in Mauss, *Techniques*, 53.

Chapter 3: Technically Speaking

1. W.F.R., "Gadget," *Notes and Queries*, 9th ser., 3 (1899): 488. For a recent edited collection of notes, correspondence, and folklore from this journal, see Edward Welch, ed., *Captain Cuttle's Mailbag: History, Folklore, and Victorian Pedantry from the Pages of "Notes and Queries"* (Astoria, N.Y.: Laboratory, 2017).
2. Archibald Sparke, "Gadget," *Notes and Queries*, 12th ser., 4 (1918): 187.
3. "Gadget," *Notes and Queries*, 12th ser., 4 (1918): 281-82.
4. A description of *Notes and Queries* by its founder, William J. Thoms, who also coined the word *folklore*. Quoted by Hannes Mandel, "Readers' Lore: Media, Literature, and the Making of Folk-lore" (PhD thesis, Princeton University, 2018), 229.
5. "Queries and Answers," *Marine Engineering* 5 (1900): 316.
6. Ralph Delahaye Paine, *The Fighting Fleets: Five Months of Active Service with the American Destroyers and their Allies in the War Zone* (Boston: Houghton Mifflin, 1918), 125; Peter Clark Macfarlane, *The Exploits of Bilge and Ma* (Boston: Little, Brown, 1919), 55; Rudyard Kipling, *Traffics and Discoveries* (New York: Doubleday, Page, 1904), 166.

7. Walter Benjamin, "The Storyteller," in *Walter Benjamin: Selected Writings, Volume 3: 1935–1938*, ed. Howard Eiland and Michael W. Jennings (Cambridge, Mass.: Belknap, 2006), 144.
8. Ludwig Wittgenstein, *Philosophical Investigations* (1953), ed. P. M. S. Hacker and Joachim Schulte, 4th ed. (Malden, Mass.: Wiley-Blackwell, 2009), § 43.
9. Toril Moi, *Revolution of the Ordinary: Literary Studies after Wittgenstein, Austin, and Cavell* (Chicago: University of Chicago Press, 2017), 33.
10. Karl Marx, *Capital: Volume One* (1867), trans. Samuel Moore and Edward Aveling, Marx/Engels Internet Archive (Marxists.org), https://www.marxists.org/archive/marx/works/1867-c1/.
11. Helen M. Rozwadowski, *Fathoming the Ocean: The Discovery and Exploration of the Deep Sea* (Cambridge, Mass.: Belknap, 2008).
12. Six months after the Battle of Manila, the shipboard wireless telegraph was an innovation that commentators immediately associated with military applications for the Philippine–American war. "Marconi and His Wireless-Telegraph System at the International Yacht Races of 1899," *Western Electrician*, October 14, 1899, 216.
13. Rozwadowski, *Fathoming the Ocean*, 21.
14. Sailors' yarns "often follow a similar pattern. First, the teller of the yarn lays the groundwork for the yarn with not only believable but mundane and picayune detail that seems unquestionable. This is followed by a believable crisis. The yarn concludes with a preposterous ending, any particular detail of which is predicated on the earlier material, leading the listener to credulity until the ultimate moment (and, for the unwary innocent, beyond)." Mary K. Bercaw Edwards, *Cannibal Old Me: Spoken Sources in Melville's Early Works* (Kent, Ohio: Kent State University Press, 2011), 45.
15. Hester Blum, *The View from the Masthead: Maritime Imagination and Antebellum American Sea Narratives*, new ed. (Chapel Hill: University of North Carolina Press, 2008), 7.
16. Robert Brown, *Spunyarn and Spindrift: A Sailor Boy's Log of a Voyage Out and Home in a China Tea-clipper* (London: Houlston and Sons, 1886), 378.
17. In fact, this passage from Brown's *Spunyarn and Spindrift* is essentially paraphrased from Melville's 1849 novel *Redburn*, where the protagonist says: "There is such an infinite number of totally new names of things to learn, that at first it seemed impossible for me to master them all. If you have ever seen a ship, you must have remarked what a thicket of ropes there are; and how they all seemed mixed and entangled together like a great skein of yarn. Now the very smallest of these ropes has its own proper name, and many of them are very lengthy, like the names of young royal princes, such as the starboard-main-top-gallant-bow-line, or the

larboard-fore-top-sail-clue-line." Herman Melville, *Redburn* (New York: Modern Library, 2002), 75.
18. Edwards, *Cannibal Old Me*, 25.
19. Edwards, 28.
20. Richard Henry Dana Jr., *The Seaman's Friend* (Boston: Charles C. Little & James Brown, and Benjamin Loring & Co., 1841), 160. For more on marlinspike seamanship, see the now-canonical handbook: Hervey Garrett Smith, *The Marlinspike Sailor* (Tuckahoe, N.Y.: J. De Graff, 1971).
21. Brown, *Spunyarn and Spindrift*, 5.
22. The earliest imaginaries of the internet portrayed a cyberspace navigated by pioneering explorers. I'd wager the nautical-to-digital influence can be attributed to a need for language to describe this navigation.
23. Adrian Johns, *Piracy: The Intellectual Property Wars from Gutenberg to Gates* (Chicago: University of Chicago Press, 2009). For another take on the sociopolitical experimentation of eighteenth-century pirates, see David Graeber, *Pirate Enlightenment, or The Real Libertalia* (New York: Farrar, Straus & Giroux, 2023). One etymology has it that ships' books were called logs because they "recorded the speed measurements made by means of a weighted chip of a tree log on the end of a reeled log line (typically 150 to 200 fathoms). The log lay dead in the water, and sailors counted the time it took the line to play out. The line was marked by different numbers of knots, or colored rags, tied at regular intervals; hence the nautical measurement sense of knot (n.)." Harper Douglas, "Etymology of log," *Online Etymology Dictionary*, accessed February 15, 2024, https://www.etymonline.com/word/log. I am indebted to John Durham Peters for several of these other connections, from his conversation with Shannon Mattern at the Center for Digital Humanities at Princeton, September 21, 2023.
24. Edwards, *Cannibal Old Me*, 24.
25. Blum, *View from the Masthead*, 5.
26. Blum, *View from the Masthead*, 41.
27. Preface to Richard Henry Dana Jr., *Two Years before the Mast and Other Voyages* (1840), ed. Thomas Philbrick (New York: Library of America, 2005), http://etext.lib.virginia.edu/toc/modeng/public/DanTwoy.html.
28. Dana, *Two Years before the Mast*.
29. For more on the concepts of neology and the novum, see Istvan Csicsery-Ronay, *The Seven Beauties of Science Fiction* (Middletown, Conn.: Wesleyan University Press, 2008).
30. James Murray, "An Appeal to the English-Speaking and English-Reading Public, to Read Books and Make Extracts for the Philological Society's New English Dictionary," April 1879, https://public.oed.com/history/archives/april-1879-appeal/.

31. Lindsay Rose Russell, *Women and Dictionary-making: Gender, Genre, and English Language Lexicography* (Cambridge: Cambridge University Press, 2018), 152.
32. According to Simon Winchester, *The Meaning of Everything: The Story of the Oxford English Dictionary*, reprint ed. (Oxford: Oxford University Press, 2004).
33. "Arrangers and subeditors assembled all relevant material for a lettered section or subsection, sorted quotation slips alphabetically and chronologically, divided words into senses and subsenses, determined the order and relation of entries, drafted definitions, and then returned their work to editors to be revised or refined. Because so much of subediting consisted in arranging, the two roles cannot always be distinguished from one another." Russell, *Women and Dictionary-making*, 158, 162.
34. Winchester, *Meaning of Everything*, 112, 99.
35. Quoted in Kory Stamper, *Word by Word: The Secret Life of Dictionaries*, reprint ed. (New York: Vintage, 2018), 173.
36. Winchester, *Meaning of Everything*, 100.
37. Henry Bradley, "Prefatory Note to Frank-law—Glass-cloth," in *Oxford English Dictionary*, vol. 4, F-G, 1910, https://public.oed.com/history/oed-editions/.
38. Russell, *Women and Dictionary-making*, 152, 153.
39. "When a lexicographer has arbitrarily decided how many 'meanings' he can conveniently recognize in the uses of a given word, he limits his entries accordingly and, after definitions of the 'meanings' in *shifted terms*, he supports them by *citations*, usually with literary authority. Lexicographical citations are keyed to the definitions, intended to exemplify a series of different 'meanings' arbitrarily selected and defined, and also to illustrate changes of meaning." John R. Firth, "A Synopsis of Linguistic Theory, 1930-1955," in *Studies in Linguistic Analysis* (Oxford: Blackwell, 1957), 1-32, at 13. While Firth popularized the thesis that words acquire meaning from their context—known as distributional semantics—with his "company it keeps" turn of phrase, the idea originally came from Zellig S. Harris, "Distributional Structure," *Word* 10, no. 2-3 (1954): 146-62, https://doi.org/10.1080/00437956.1954.11659520.
40. For more on Masterman's founding of the Cambridge Language Research Unit and her creation with Karen Spärck Jones of a computer-based thesaurus, see Michael Gavin, "Vector Semantics, William Empson, and the Study of Ambiguity," *Critical Inquiry* 44, no. 4 (2018): 641-73, https://doi.org/10.1086/698174. Both Masterman and Alan Turing were students of Wittgenstein's, with the former transcribing the lectures that would become the *Blue and Brown Books*.

41. Lydia H. Liu, "Wittgenstein in the Machine," *Critical Inquiry* 47, no. 3 (2021): 425-55, at 450, https://doi.org/10.1086/713551.
42. Daniel Jurafsky and James H. Martin, *Speech and Language Processing: An Introduction to Natural Language Processing, Computational Linguistics, and Speech Recognition*, 3rd ed. draft, 2024, 130. Jurafsky and Martin cite many other NLP researchers who trace their work back to Firth's, and by extension Wittgenstein's, thinking. Lydia Liu asks, "I wonder sometimes if the AI scientist's Wittgenstein is the same thinker as the philosopher's Wittgenstein." Liu excavates the many ways AI researchers from the 1970s on have drawn on Wittgenstein, whether the comparison is apt or not. "Wittgenstein's philosophy of language is so closely bound up with the semantic networks of the computer from the mid-1950s down to the present that we can no longer turn a blind eye to its embodiment in the AI machine. But what does it mean for AI practitioners to engage with Wittgenstein?" Liu, "Wittgenstein in the Machine," 428.
43. "Languages are systems of signs, i.e., pairings of form and meaning. But the training data for LMs [language models] is only form; they do not have access to meaning." Emily Bender et al., "On the Dangers of Stochastic Parrots: Can Language Models Be Too Big?" in *FAccT '21*, 2021-03-03/2021-03-10, citing Emily M. Bender and Alexander Koller, "Climbing towards NLU: On Meaning, Form, and Understanding in the Age of Data," in *Proceedings of the 58th Annual Meeting of the Association for Computational Linguistics*, ed. Dan Jurafsky et al. (Kerrville, Tex.: Association for Computational Linguistics, 2020), 5185-98, https://doi.org/10.18653/v1/2020.acl-main.463.
44. Spoken by the titular character in the novel *Ahab's Wife*. Sena Jeter Naslund, *Ahab's Wife, or The Star-gazer* (1999; reprint, New York: William Morrow Paperbacks, 2005).
45. "Gadget, n.," *Oxford English Dictionary*, accessed May 12, 2018, http://dictionary.oed.com/.
46. Michael Quinion, *Port Out, Starboard Home and Other Language Myths* (London: Penguin, 2005).
47. *Webster's Deluxe Unabridged Dictionary*, 2nd ed. (New York: New World Dictionaries/Simon and Schuster, 1979).
48. E. Hospitelier, *Vocabulaire Français-Anglais-Allemand: Technique, Industriel, et Commercial* (Paris: A. Lahure, 1909). The word was not included in the 1900 first edition. Armand Antoine Agénor de Gramont, *The Aviator's Pocket Dictionary and Table-book, French-English and English-French* (New York: Bretano's, 1918). For apparatus theories of cinema, see Philip Rosen, ed., *Narrative, Apparatus, Ideology: A Film Theory Reader* (New York: Columbia University Press, 1986).

49. Herman Melville, *Moby-Dick, or The White Whale* (New York: Dodd, Mead, 1923), 528, 529.
50. Other banished words that were too technically specialized included *calkin-iron* and *shrouds*, both of which he found in the poetry of Dryden. Samuel Johnson, *Lives of the Poets* (1779-81), Project Gutenberg e-book, https://www.gutenberg.org/.
51. Alfred Henry Alston, *Seamanship and Its Associated Duties in the Royal Navy* (London: Routledge, Warne, & Routledge, 1860), 119-20.
52. Clifford W. Ashley, *The Ashley Book of Knots* (New York: Doubleday, 1944), 421.
53. Vita Sackville-West, *Country Notes* (New York: Harper & Bros., 1940), 197.
54. Sackville-West, 196, 198.
55. Sources for the preceding uses of *gadget* are, in order: Homer Saint-Gaudens, "Man and the Machine," *Metropolitan Magazine* 27, no. 6 (1908); R. Stanley Edwards, "Aeronitis," *Aerial Age Weekly*, July 1919; "Selling to Toledo: A Big Market that Should Interest Sellers of Many Lines of Goods," *Toledo City Journal* 4, no. 5 (1919); Helen Louise Walker, "A Gadget a Day," *Picture Play Magazine*, November 1937; Walter Brooks, "Once Over Lightly: Armchair Musings of Our Philosophic Observer: Gadget Theory," *Scribner's Commentator* 7, no. 4 (1940): 42; "Actor vs. Gadget: Clifton Webb Discovers Seven Interesting Uses for a New Gimmick, None of Them Right," *Life Magazine*, January 1949; *A. & P. Tea Co. v. Supermarket Corp.*, December 1950; Jean Sala Breitenstein, *Jamco, Inc. v. Theodor F. Carlson*, 1959; P. L. Giovacchini, "On Gadgets," *Psychoanalytic Quarterly* 28 (1959): 330-41; Jack Kerouac, *Visions of Cody* (New York: McGraw-Hill, 1972); John Updike, *The Music School* (New York: Knopf, 1966); Herbert Marcuse, *Eros and Civilization: A Philosophical Inquiry into Freud* (Boston: Beacon, 1955); Don DeLillo, *Ratner's Star* (New York: Knopf, 1976); Harry Hurt III, "Moonwalk," *Newsweek*, September 18, 1989, 48-49; D. T. Dingle, "Ripped Off!," *Money* 20, no. 8 (1991): 96; Piers Anthony, *A Spell for Chameleon* (New York: Ballantine, 1977); Jeffrey A. Tannenbaum, "Consumer Electronics May Lack Blockbuster, but Not New Gadgets," *Wall Street Journal*, January 1988; "Always in the Loop: Get Spiegel Online's Widgets and Gadgets for Netvibes and iGoogle," *Spiegel*, August 17, 2009; Marshall McLuhan, *The Mechanical Bride: Folklore of Industrial Man* (New York: Vanguard, 1951), 32.
56. "Egregious, adj.," and "actually, adv.," *Oxford English Dictionary*, accessed October 1, 2021, http://dictionary.oed.com/.
57. The BC-221 frequency meter, for example, is a gadget detailed in the famous amateur radio magazine *QST*. "Technical Topics—Re: Half-Wave Filters," *QST: American Radio Relay League* 34 (1950): 66.

58. Henry Schlesinger, *The Battery: How Portable Power Sparked a Technological Revolution* (New York: HarperCollins, 2011), Kindle loc. 3545.
59. Igo writes of the ways privacy discourse shifted to accommodate ever smaller devices that could surreptitiously record conversations in mid-1960s America. This included futuristic spy gadgets like "miniature transmitters, radio pills, two-way mirrors, electronic eyes, hidden television-eye monitoring, closed-circuit television, infrared film, . . . micro-miniature tape recorders, electrically conductive paint," and more. Sarah Igo, *The Known Citizen: A History of Privacy in Modern America* (Cambridge, Mass.: Harvard University Press, 2018), 168, 226.
60. Haraway, "Cyborg Manifesto," 12–13.
61. Samuel R. Delany, "Black to the Future: Interviews with Samuel R. Delany, Greg Tate, and Tricia Rose," in *Flame Wars: The Discourse of Cyberculture*, ed. Mark Dery (Durham, N.C.: Duke University Press, 1994), 179–222, at 192. On the link between audio technology and modern Black culture, see Alexander G. Weheliye, *Phonographies: Grooves in Sonic Afro-modernity* (Durham, N.C.: Duke University Press, 2005), https://doi.org/10.1215/9780822386933.
62. Clive Thompson, "You Know Who's Really Addicted to Their Phones? The Olds," *Wired*, March 27, 2018.
63. Jody Rosen, "In Praise of the Humble Knot," *New York Times*, September 17, 2014.
64. Harold Augustin Calahan, *Gadgets and Wrinkles: A Compendium of Man's Ingenuity at Sea* (New York: Macmillan, 1938), ix. Calahan dedicates the book to his father, inventor Edward Augustin Calahan, "whose great gadgets, the stock ticker, the messenger boy call box, and the multiplex telegraph, annihilated time and space." The application of the term to electrical devices was highly uncommon at the time.
65. Calahan, 3. A related distinction between *technique* and *wrinkle* is made along the lines of complexity in "Wrinkles and Gadgets," *Watson's Microscope Record*, no. 22–28 (1931), 23: "If technique is, as Mark Twain suggested, doing something simple in a complicated way, then a wrinkle suggests the reverse of this; that is, a simple method of doing something complicated."
66. Melville, *Moby-Dick*, 423.

Chapter 4: The Custody of Automatism

1. Philly Community Wireless, "About," accessed March 24, 2024, https://phillycommunitywireless.org/. Allan Gomez et al., "Alternative Infrastructures for Digital Equity: Community-Based Internet Access," in *Critical Infrastructure Studies and Digital Humanities*, ed. Alan Liu, Urszula

Pawlicka-Deger, and James Smithies (Minneapolis: University of Minnesota Press, forthcoming).
2. Greta Byrum, "Building the People's Internet," *Urban Omnibus*, October 2, 2019.
3. PCW learned from the rich history of past community technology efforts in Philadelphia. These include Wireless Philadelphia, one of the country's first municipal Wi-Fi providers that unfortunately dissolved soon after it was formed in 2004; Prometheus, a collective that built and advocated for independent, low-power radio and other wireless technologies throughout the 2000s; and Philly Mesh, an experimental group that worked toward interoperability with mesh networks in other cities in the mid-2010s. Joshua Breitbart, Naveen Lakshmipathy, and Sascha D. Meinrath, "The Philadelphia Story: Learning from a Municipal Wireless Pioneer" (Washington, D.C.: New America Foundation, December 2007). Christina Dunbar-Hester, *Low Power to the People: Pirates, Protest, and Politics in FM Radio Activism* (Cambridge, Mass.: MIT Press, 2014). Philly Mesh, https://phillymesh.net/, 2013.
4. According to one study, "A 2019 survey by the School District of Philadelphia found that only 45% of students in grades three through five accessed the internet from a computer at home, compared with 56% in grades six through eight, and 58% for high school students." Christian Hetrick and Dylan Purcell, "Thousands of Philly Students Are Stuck at Home with No Computer or Internet after Coronavirus Closed Schools," *Philadelphia Inquirer*, April 3, 2020.
5. Alvaro Sanchez, "Toward Digital Inclusion: Broadband Access in the Third Federal Reserve District," Federal Reserve Bank of Philadelphia, March 2020.
6. See, for example, two Philadelphia maps based on 2018 data from the U.S. Census Bureau's American Community Survey: a map of disconnection rates and a map of race and ethnicity. Dylan Purcell, "Broadband Internet Access Lags in Many Philadelphia Neighborhoods," Datawrapper, 2021; "Race and Ethnicity in Philadelphia, Pennsylvania," Statistical Atlas, September 2018, https://statisticalatlas.com/place/Pennsylvania/Philadelphia/Race-and-Ethnicity#data-map/neighborhood/white.
7. "Equitable Internet Initiative," Detroit Community Technology Project, accessed April 6, 2024, https://detroitcommunitytech.org/eii. For an overview of the foundational community network projects in Europe, see Leandro Navarro et al., "Network Infrastructure as Commons: Report on Existing Community Networks and Their Organization," cofunded by the Horizon 2020 program of the European Union, September 2016.

8. Diana J. Nucera, "Teaching Community Technology Handbook" (Detroit Community Technology Project, 2016), 15.
9. Byrum, "Building the People's Internet."
10. As Hobart and Kneese write in their introduction to an important issue of *Social Text* devoted to "Radical Care," such forms of care have historically come to the fore "when institutions and infrastructures break down, fail, or neglect. Reciprocity and attentiveness to the inequitable dynamics that characterize our current social landscape represent the kind of care that can radically remake worlds that exceed those offered by the neoliberal or postneoliberal state, which has proved inadequate in its dispensation of care." Hiʻilei Julia Kawehipuaakahaopulani Hobart and Tamara Kneese, "Radical Care: Survival Strategies for Uncertain Times," *Social Text* 38, no. 1 (2020): 1–16, at 3, https://doi.org/10.1215/01642472-7971067.
11. David Kravets, "U.N. Report Declares Internet Access a Human Right," *Wired*, June 3, 2011.
12. David Rosen, "Cities Struggle to End the Urban Digital Divide," *Progressive*, September 2021, https://progressive.org/api/content/be46bbdc-16ef-11ec-bb5c-1244d5f7c7c6/. These community technologies respond to the prevailing mode of internet access in the United States: even though there is broad agreement that the internet is a public good, neoliberal policies have only privileged private models of internet service, financializing the commons and withholding access and ownership from exploited communities.
13. Mimi Onuoha, "The Cloth in the Cable," exhibition presented at Australian Centre for Contemporary Art, 2022, https://mimionuoha.com/in-the-cable.
14. For examples of this self-help genre of digital minimalism, see Tara Brabazon, *Digital Dieting: From Information Obesity to Intellectual Fitness* (London: Routledge, 2013); James Clear, *Atomic Habits: An Easy and Proven Way to Build Good Habits and Break Bad Ones* (New York: Avery, 2018); David M. Levy, *Mindful Tech: How to Bring Balance to Our Digital Lives*, reprint ed. (New Haven, Conn.: Yale University Press, 2017); Cal Newport, *Digital Minimalism: Choosing a Focused Life in a Noisy World* (New York: Portfolio, 2019).
15. "For Technologists: Review Our Principles," Center for Humane Technology, accessed October 19, 2021, https://www.humanetech.com/technologists#principles.
16. Williams continues: "In many cases, this rejection occurred on the basis of philosophical or cosmological disagreements with the old packages. This has, of course, had many great benefits. Yet by rejecting entire packages of constraint, we've also rejected those constraints that were actually useful for our purposes. When you dismantle existing boundaries in your

environment, it frees you from their limitations, but it requires you to bring your own boundaries where you didn't have to before. Sometimes, taking on this additional self-regulatory burden is totally worth it. Other times, though, the cost is too high." James Williams, *Stand Out of Our Light: Freedom and Resistance in the Attention Economy*, reprint ed. (Cambridge: Cambridge University Press, 2018), 21-22.

17. Crawford continues: "This created a vacuum of cultural authority that has been filled, opportunistically, with attentional landscapes that get installed by whatever 'choice architect' brings the most energy to the task—usually because it sees the profit potential." Matthew B. Crawford, *The World beyond Your Head: On Becoming an Individual in an Age of Distraction*, reprint ed. (New York: Farrar, Straus & Giroux, 2016), 41.

18. Siegfried Kracauer, "Boredom," in *The Mass Ornament: Weimar Essays*, ed. Thomas Y. Levin (Cambridge, Mass.: Harvard University Press, 1995), 331-36, at 332-33, 334.

19. "Allocating responsibility for social problems to individuals is part of the neoliberal agenda that characterise information policies. With neoliberalism and deregulation, individuals are allocated more responsibility for managing areas such as health and lifestyle, employment, crime prevention, workplace safety and online risks." Syvertsen, *Digital Detox*, 8.

20. Laura Portwood-Stacer, "Media Refusal and Conspicuous Nonconsumption: The Performative and Political Dimensions of Facebook Abstention," *New Media and Society* 15, no. 7 (2013): 1041-57, at 1041, https://doi.org/10.1177/1461444812465139. See also Karppi, *Disconnect*.

21. Jenny Odell, *How to Do Nothing: Resisting the Attention Economy* (Brooklyn, N.Y.: Melville House, 2019).

22. Megan Ward, "Your Digital Detox Isn't as Radical as You Think," *Washington Post*, July 30, 2017. Merchant has provided the timely reminder that nineteenth-century Luddism was a highly technoliterate movement of skilled technicians who originally aspired to benefit from the leisure afforded by machines that automated their labor. When the designs of factory owners proved otherwise—they simply fired the workers rather than sharing the value produced by the machines—the Luddites began organizing. Brian Merchant, *Blood in the Machine* (Boston: Little, Brown, 2023).

23. Ticona conducts a comparative study of how workers in different sectors manage the "entrance of mobile phones into the workplace . . . within a marketplace where jobs are increasingly uncertain and insecure." She finds that "service workers deployed strategies of everyday resistance in concert with their ICTs to gain a feeling of autonomy within the power structures of their workplaces. The knowledge workers deployed strategies of inaccessibility to resist the work-extending affordances of their devices

and decouple from work which threatened to colonize too much of their lives. Both service and knowledge workers deploy strategies that may obscure the institutional sources of their problems by overindividualizing risk and responsibility." Julia Ticona, "Strategies of Control: Workers' Use of ICTs to Shape Knowledge and Service Work," *Information, Communication, and Society* 18, no. 5 (2015): 509–23, at 509, https://doi.org/10.1080/1369118X.2015.1012531. On pretrial sentencing, see especially the Mapping Pretrial Injustice Project, a collaboration between the Movement Alliance Project and MediaJustice, at https://pretrialrisk.com/.
24. Ruha Benjamin, *Race after Technology: Abolitionist Tools for the New Jim Code* (Cambridge: Polity, 2019), 15–16, 17.
25. *Offline Is the New Luxury*, dir. Bregtje van der Haak, 2016.
26. "What's needed, instead of a pretense to purity that is impossible in the actually existing world, is something else. We need to shape practices of responsibility and memory for our placement in relation to the past, our implication in the present, and our potential creation of distant futures." Alexis Shotwell, *Against Purity: Living Ethically in Compromised Times* (Minneapolis: University of Minnesota Press, 2016), 8.
27. Mauss, *Techniques*, 152.
28. For some of the most influential accounts, see Nicholas G. Carr, *The Shallows: What the Internet Is Doing to Our Brains* (New York: Norton, 2010; reissued in 2020 for a tenth-anniversary edition); Paul North, *The Problem of Distraction* (Stanford, Calif.: Stanford University Press, 2012); Alex Soojung-Kim Pang, *The Distraction Addiction: Getting the Information You Need and the Communication You Want, without Enraging Your Family, Annoying Your Colleagues, and Destroying Your Soul* (Boston: Little, Brown, 2013); Dominic Pettman, *Infinite Distraction* (Cambridge: Polity, 2015).
29. Dennis Yi Tenen, *Literary Theory for Robots: How Computers Learned to Write* (New York: Norton, 2024).
30. For several versions of this parable, see Dionysius Lardner, *The Steam Engine Familiarly Explained and Illustrated; with an Historical Sketch of Its Invention and Progressive Improvement; Its Applications to Navigation and Railways* (London: Taylor, Walton, and Maberly, 1851), 53; Andrew Carnegie, *James Watt* (New York: Doubleday, Page, 1905); D. C. Beard, *The Outdoor Handy Book: For Playground, Field, and Forest* (New York: Scribner, 1914); Marshall McLuhan, *Understanding Me: Lectures and Interviews*, ed. David Staines and Stephanie McLuhan (Cambridge, Mass.: MIT Press, 2003), 292–95. Somehow, none of the sources I've come across think to remark on the conventionality of child labor in industrial England!
31. Aaron Bastani, *Fully Automated Luxury Communism: A Manifesto*, reprint ed. (London: Verso, 2020). Nick Srnicek and Alex Williams, *Inventing the Future: Postcapitalism and a World without Work* (London: Verso, 2015).

32. "Crossmatch [an employee analytics vendor for companies like KFC, Wendy's, and RiteAid] touts this system as preventing 'tardy arrivals, buddy punching, lollygagging, extended breaks and early departures, inventory shrink, unauthorized discounts and returns, and fraudulent gift card transactions;' anyone who's ever worked retail will understand this list as an almost perfect, point-for-point recitation of the tacit measures employees have always taken to compensate themselves for having to put up with abusive bosses, shitty pay and intolerable working conditions. For that matter, 'buddy punching'—the act of clocking in a friend who's late for work, possibly because they've had to take a child to daycare or a sick parent to the doctor—is just the kind of small act of solidarity that might save someone their job in the harsh, zero-tolerance climate of contemporary work. But these are the tactics such oversight systems are expressly designed to eliminate." Adam Greenfield, *Radical Technologies: The Design of Everyday Life* (London: Verso, 2017), 196, 198. For more on worker surveillance and "strategies of everyday resistance," see Ticona, "Strategies of Control."
33. Benjamin H. Bratton, *The Stack: On Software and Sovereignty* (Cambridge, Mass.: MIT Press, 2016), 255.
34. Bender et al., "On the Dangers of Stochastic Parrots."
35. Kyle Wiggers, "Text Autocompletion Systems Aim to Ease Our Lives, but There Are Risks," *VentureBeat*, January 11, 2022.
36. Gish Jen, *The Resisters* (New York: Vintage, 2020), 8-9.
37. Already in the 1890s, Alfred Espinas excluded "unconscious practices" from his definition of *technique*, a term that took on a moral valence in emphasizing thoughtful action over passive habit: "It would however be most profitable if we could use that word, as the Greeks did, for conscious and well thought-out practices, to a certain extent opposed to simple practices, which develop spontaneously, prior to any analysis. For it is fully developed arts, not unconscious practices, which gave birth to the science we are dealing with, and generate Technology." Espinas, "Origins of Technology," 46-47; Espinas, *Les origines de la technologie*, 7-8.
38. William James, *The Principles of Psychology* (New York: Holt, 1890), 122. For a review of the scant philosophical scholarship on "unreflective skillful action," see Erik Rietveld, "McDowell and Dreyfus on Unreflective Action," *Inquiry* 53, no. 2 (2010): 183-207, https://doi.org/10.1080/00201741003612203.
39. Julian E. Orr, *Talking about Machines: An Ethnography of a Modern Job* (Ithaca, N.Y.: ILR Press, 1996). Sennett, *Craftsman*. Pamela H. Smith, *From Lived Experience to the Written Word: Reconstructing Practical Knowledge in the Early Modern World* (Chicago: University of Chicago Press, 2022).
40. Rose comes to a different conclusion than mine about the expertise of the user. Ellen Rose, *User Error: Resisting Computer Culture* (Toronto: Between the Lines, 2003), 1.

41. Dery, "Black to the Future," 192.
42. Rose, *User Error*, 1.
43. For a broader take on the consequences of hyperspecialization and knowledge, see Elijah Millgram, *The Great Endarkenment: Philosophy for an Age of Hyperspecialization* (Oxford: Oxford University Press, 2015).
44. Kylie Foy, "A Method to Interpret AI Might Not Be So Interpretable After All," *MIT News*, October 16, 2023. Reporting on Ho Chit Siu, Kevin Leahy, and Makai Mann, "STL: Surprisingly Tricky Logic (for System Validation)," arXiv, May 26, 2023, https://doi.org/10.48550/arXiv.2305.17258.
45. A frequently cited passage from Alfred North Whitehead comes to mind: "It is a profoundly erroneous truism, repeated by all copy-books and by eminent people when they are making speeches, that we should cultivate the habit of thinking of what we are doing. The precise opposite is the case. Civilization advances by extending the number of important operations which we can perform without thinking about them. Operations of thought are like cavalry charges in a battle—they are strictly limited in number, they require fresh horses, and must only be made at decisive moments." This quotation is often deployed as a definition of technology, but it originally appears in a discussion of the value of symbols as shorthand in mathematical equations and logical proofs: Alfred North Whitehead, *An Introduction to Mathematics* (1958), Project Gutenberg e-book, https://www.gutenberg.org/.
46. For a brilliant catalog of the ways science fiction has influenced user interface design, see Nathan Shedroff and Christopher Noessel, *Make It So: Interaction Design Lessons from Science Fiction* (Brooklyn: Rosenfeld Media, 2012).
47. Tresch describes a 1959 display at the Musee de l'homme in Paris, in the Hall of Arts and Techniques, which was then under the direction of André Leroi-Gourhan. The display included photographs of people at work with various tools alongside stylized graphical outlines of those individuals' gestures in using the tools. Leroi-Gourhan's "approach shifted attention away from objects toward the physical actions they embodied and enhanced. He saw technical objects as realizing certain 'tendencies,' externalizing gestures previously performed by the body alone." I find this speculative approach fascinating: removing the tool from the picture and foregrounding gesture as a means of indicating what the user knows. John Tresch, "Leroi-Gourhan's Hall of Gestures," in *Energies in the Arts*, ed. Douglas Kahn (Cambridge, Mass.: MIT Press, 2019), 193–238, at 223.
48. Michael Polanyi, *The Tacit Dimension* (Garden City, N.Y.: Doubleday, 1967), 4. Similarly, Stanley argues that "know-how" is a form of knowledge: "everyone who discusses skilled action, from [Gilbert] Ryle forwards,

agrees that skilled action requires knowledge how. The debate has been about the nature of knowledge how. I have argued that skilled action is action guided by knowledge how, and that knowing how to do something amounts to knowing a fact. Skilled action is action guided by knowledge of facts." Jason Stanley, *Know How,* reprint ed. (Oxford: Oxford University Press, 2013), 175, citing Gilbert Ryle, "Knowing How and Knowing That," in *The Concept of the Mind* (London: Penguin, 1949).

49. A number of laboratory studies conducted by science and technology studies scholars explore the folk theories possessed by users of social media platforms: Motahhare Eslami et al., "First I 'Like' It, Then I Hide It: Folk Theories of Social Feeds," in *Proceedings of the 2016 CHI Conference on Human Factors in Computing Systems,* CHI '16 (New York: Association for Computing Machinery, 2016), 2371–82, https://doi.org/10.1145/2858036.2858494; Megan French and Jeff Hancock, "What's the Folk Theory? Reasoning about Cyber-social Systems," *SSRN Electronic Journal,* 2017, https://doi.org/10.2139/ssrn.2910571; Nadia Karizat et al., "Algorithmic Folk Theories and Identity: How TikTok Users Co-produce Knowledge of Identity and Engage in Algorithmic Resistance," *Proceedings of the ACM on Human-Computer Interaction* 5, no. CSCW2 (2021): article 305, https://doi.org/10.1145/3476046.

50. "Inscrutability is a difficult problem in practice, in no small part because software systems have become incredibly powerful. As policy and engineering practice evolves to deal with the newfound importance of software systems, it is critical to avoid attributing lack of understanding of computer systems to their massive technical complexity. Rather, we must use the context and history of systems to build a more complete understanding that avoids the fallacy of inscrutability." Joshua A. Kroll, "The Fallacy of Inscrutability," *Philosophical Transactions of the Royal Society A: Mathematical, Physical and Engineering Sciences* 376, no. 2133 (2018), https://doi.org/10.1098/rsta.2018.0084.

51. Michel de Certeau, *The Practice of Everyday Life* (Berkeley: University of California Press, 1988), xi, xiii.

52. "Moreover, these stories represented a discourse in which gender was invisible. Historians did not consider it relevant in settings where women were absent, thus reinforcing the view that men had no gender." Nelly Oudshoorn and Trevor Pinch, introduction to *How Users Matter: The Co-construction of Users and Technology,* ed. Nelly Oudshoorn and Trevor Pinch (Cambridge, Mass.: MIT Press, 2005), 1–28, at 4, citing in particular Judy Wajcman, *Feminism Confronts Technology* (University Park: Pennsylvania State University Press, 1991); Nina E. Lerman, Arwen Palmer Mohun, and Ruth Oldenziel, "Versatile Tools: Gender Analysis and the History of

Technology," *Technology and Culture* 38, no. 1 (1997): 1–8, https://doi.org/10.2307/3106781.
53. Molly Wood, "In 2019, Your Smartphone Will No Longer Be King," Marketplace Tech (blog), January 1, 2019. Kara Swisher, "No More Phones and Other Tech Predictions for the Next Decade," *New York Times*, December 31, 2019. Lauren Goode, "The Future Smartphone: More Folds, Less Phone, a Whole Lot of AI," *Wired*, March 15, 2023.
54. For more on solar mesh nodes, see https://docs.phillycommunitywireless.org/en/latest/installations/solarmesh/.
55. David Nield, "How Amazon Sidewalk Works—And Why You May Want to Turn It Off," *Wired*, May 11, 2021.
56. Shannon Hall, "As SpaceX Launches 60 Starlink Satellites, Scientists See Threat to 'Astronomy Itself,'" *New York Times*, November 11, 2019.
57. Kevin Roose, "Maybe There's a Use for Crypto After All," *New York Times*, February 6, 2022.

Epilogue: Reclaiming Technique

1. A legal think tank at Georgetown University sidestepped this morass when they released a statement on their refusal to use the term *AI* or even *machine learning* in their research: "Starting today, the Georgetown Center on Privacy & Technology Law will stop using the terms 'artificial intelligence,' 'AI,' and 'machine learning' in our work to expose and mitigate the harms of digital technologies in the lives of individuals and communities." Emily Tucker, "Artifice and Intelligence," Center on Privacy and Technology at Georgetown Law, March 8, 2022.
2. Samanth Subramanian, "AI and the End of the Human Writer," *New Republic*, April 22, 2024.
3. "It's widely believed that Jaan Tallinn, the wealthy long-termer who co-founded the most prominent centers for the study of AI safety, has made dismissive noises about climate change because he thinks that it pales in comparison with far-future unknown unknowns like risks from AI. The technology historian David C. Brock calls these fears 'wishful worries'— that is, 'problems that it would be nice to have, in contrast to the actual agonies of the present.'" Bruce Schneier and Nathan Sanders, "The AI Wars Have Three Factions, and They All Crave Power," *New York Times*, September 28, 2023.
4. For just a sampling of this pre-ChatGPT, critical work on AI from the perspective of the humanities, see Meredith Broussard, *Artificial Unintelligence: How Computers Misunderstand the World* (Cambridge, Mass.: MIT Press, 2018); Lise Jaillant, ed., *Archives, Access, and Artificial Intelligence:*

Working with Born-Digital and Digitized Archival Collections (Bielefeld, Germany: Bielefeld University Press, 2022); Orit Halpern et al., "Surplus Data: An Introduction," *Critical Inquiry* 48, no. 2 (2022): 197-210, https://doi.org/10.1086/717320; Lauren Klein, "Are Large Language Models Our Limit Case?," *Startwords*, no. 3 (2022), https://doi.org/10.5281/zenodo.6567985; Meredith Martin, "AI Off the Rails: Emily M. Bender on the Everything in the Whole Wide World Benchmark," Critical AI@Rutgers, April 15, 2022; Laura D. Tyson and John Zysman, "Automation, AI, and Work," *Daedalus* 151, no. 2 (2022): 256-71; John Tasioulas, "The Role of the Arts and Humanities in Thinking about Artificial Intelligence (AI)," Ada Lovelace Institute blog, June 14, 2021; Amy A. Winecoff and Elizabeth Anne Watkins, "Artificial Concepts of Artificial Intelligence: Institutional Compliance and Resistance in AI Startups," in *Proceedings of the 2022 AAAI/ACM Conference on AI, Ethics, and Society*, 2022, 788-99, https://doi.org/10.1145/3514094.3534138.

5. On the reasons why OpenAI's ChatGPT (and metaphor of a chat interface) have become "a common shorthand for the broader sociotechnical condition" of AI, see Matthew Kirschenbaum and Rita Raley, "AI and the University as a Service," *PMLA* 139, no. 3 (2024): 504-15, https://doi.org/10.1632/S003081292400052X.

6. Ethan Mollick, "I, Cyborg: Using Co-intelligence," One Useful Thing (blog), March 14, 2024, citing Anson Ho et al., "Algorithmic Progress in Language Models," arXiv, March 9, 2024, https://doi.org/10.48550/arXiv.2403.05812.

7. At the time of writing, the forward-thinking University of Minnesota Press offers a book contract clause that forbids third parties from using the author's work to train generative AI. I gratefully added that clause to my contract for this book. So it's worth pointing out here that if this text does find its way into an LLM, it was without my consent.

8. Jen, *Resisters*. IMF research projects that "almost 40 percent of global employment is exposed to AI." Mauro Cazzaniga et al., "Gen-AI: Artificial Intelligence and the Future of Work," International Monetary Fund, January 2024. On the weird as an aesthetic category best suited to thinking about AI and culture, see Erik Davis, "The Weird and the Banal," Burning Shore (Substack), April 24, 2023.

9. Consider an AI-mediated exchange that's already possible. You offend someone by text message. An AI watching the conversation recognizes that offense may have been taken, notifies you, and generates a perfectly worded apology. If a service like this becomes so widespread that your interlocutor suspects its use, will your apology mean anything, whether you used the AI or not? Simon DeDeo, "Simon DeDeo with Arthur Spirling (topic: models)," LLM Forum, Princeton University, February 21, 2024.

10. Mauss, *Techniques*, 53.
11. "Because every photograph and artwork archives a loss and, in archiving it, keeps it as loss, there is in every photograph and artwork at once an imminence of death and a delaying of it, a time—a deferral, a lag—between the verdict, 'this too will die,' 'this too owes itself to death,' and the carrying out of the verdict." Eduardo Cadava, *Paper Graveyards* (Cambridge, Mass.: MIT Press, 2021), 57. See also Roland Barthes, *Camera Lucida: Reflections on Photography* (New York: Hill & Wang, 1982); Tim Carpenter, *To Photograph Is to Learn How to Die: An Essay with Digressions* (Los Angeles: Ice Plant, 2022).
12. The original painting, from which the RCA logo is cropped, depicts the dog and gramophone sitting atop a coffin. Jonathan Sterne, *The Audible Past: Cultural Origins of Sound Reproduction* (Durham, N.C.: Duke University Press, 2003), 301–3.
13. Rebecca Carballo, "Using AI to Talk to the Dead," *New York Times*, December 11, 2023. Todd Nelson, "A New Medium for Communicating with the Dead: AI and Chatbots," *Minnesota Star Tribune*, July 6, 2023. Another start-up brings back famous artists from the dead. Kit Eaton, "Hello Dalí: Museum Exhibit Uses AI Tech for a Surreal Twist to Let Visitors 'Talk' to the Artist," MSN, April 11, 2024.
14. Schiffer, "Cultural Imperatives." See also Sterne, describing the cultural origins of sound reproduction technologies like the phonograph in "wishes that people grafted on to [those] technologies—wishes that became programs for innovation and use." Sterne, *Audible Past*, 8.
15. Timothy Morton, *Hyperobjects: Philosophy and Ecology after the End of the World* (Minneapolis: University of Minnesota Press, 2013).
16. Critics of AI, in the view of this manifesto, aren't merely standing in the way of progress. They are committing murder: "We believe any deceleration of AI will cost lives. Deaths that were preventable by the AI that was prevented from existing is a form of murder." "The Techno-optimist Manifesto," Andreessen Horowitz (blog), October 2023.
17. Catherine D'Ignazio and Lauren F. Klein, *Data Feminism* (Cambridge, Mass.: MIT Press, 2020), especially chapter 1.
18. Matthew Kirschenbaum, "Prepare for the Textpocalypse," *Atlantic*, March 8, 2023.
19. Maria del Rio-Chanona, Nadzeya Laurentsyeva, and Johannes Wachs, "Are Large Language Models a Threat to Digital Public Goods? Evidence from Activity on Stack Overflow," arXiv, July 14, 2023, https://doi.org/10.48550/arXiv.2307.07367.
20. Maggie Appleton, "The Expanding Dark Forest and Generative AI," Maggie Appleton (blog), March 2023.

21. This is of course not an endorsement of data poisoning, but rather a reference to the ways that artists are using the technique to protect their creative works from being ingested into generative AI training data without permission. See Melissa Heikkilä, "This New Data Poisoning Tool Lets Artists Fight Back against Generative AI," *MIT Technology Review*, October 23, 2023.
22. On community intranet, see Community Tech NY's Portable Network Kits (https://www.communitytechny.org/portable-network-kits) and iNethi's local servers (https://www.inethi.org.za/about/). On the use of Gopher in the 2020s, see Project Gemini (https://geminiprotocol.net/). On long range, low-power messaging, see Meshtastic (https://meshtastic.org/).
23. Ilia Shumailov et al., "The Curse of Recursion: Training on Generated Data Makes Models Forget," arXiv, submitted May 27, 2023, revised April 14, 2024, https://doi.org/10.48550/arXiv.2305.17493. This "is a phenomenon that Jathan Sadowski calls 'Habsburg AI,' where 'a system that is so heavily trained on the outputs of other generative AIs that it becomes an inbred mutant, likely with exaggerated, grotesque features.' In reality, a Habsburg AI will be one that is increasingly more generic and empty, normalized into a slop of anodyne business-speak as its models are trained on increasingly-identical content." Edward Zitron, "Are We Watching the Internet Die?," Where's Your Ed At (blog), March 11, 2024.

INDEX

2001: A Space Odyssey, 7, 95

addiction: habit framed as, 76, 78, 80, 119n34; users as, 66, 75-76, 85-86, 119n34
aërials (antennae), early forms of, 1-4
aesthetic acts and forms, 9-10, 33, 35, 52, 111n30
affordances, x, xi, 11, 46, 97, 112n39
Against Purity: Living Ethically in Compromised Times (Shotwell), 80
agency: and gadgets, 65-66; lack of attributed to smartphone users, 77-78; of users, 9, 17, 89; of women innovators, 89
agriculture, 28-31; and calendars, 40; farm plots, shape of, 22, 30; plows, medieval, 9, 30-31
algorithms, xi, xii, 57, 74-76; "algorithmic imaginary" and "conscious clicking," 34; pretrial sentencing, 79, 134n23. *See also* artificial intelligence (AI)
"American walking fashions," 25
Annales d'histoire économique et sociale, 19-21, 24
Annales school of historiography, 21, 23-24, 27
Année Sociologique, 32
antedating, 56, 58
antennae: early forms of, 1-4; PCW project, 72, 92-93

Appleton, Maggie, 99
applied sciences, 24, 41, 116n57
archaeology, 35-36, 112-13n40
Aristotle, 36
art, 33, 40, 81, 111n31
artificial intelligence (AI): conversational, 90; economic imperative of, 97-98; harms of, 76, 110-11n29, 138n1, 138n3; as human augmentation, 97; interpersonal exchanges affected by, 83, 96, 139n9; interpretability problem, 86-89; large language models (LLM), 83, 96, 128n43; model collapse, 100; natural language processing (NLP), 57-58, 61, 128n42; novelty used to outsmart, 99-100; polysemy of meanings, 61, 95, 116n57; power dynamics of, 88-89; predictive text systems, 83; and user intuition, 89; worldwide rollout, 95-96. *See also* algorithms
artisans, 16, 24; nonspecialist and skilled, 85-86
attention, 76-77; as central technique for public discourse, 80-81; and cultural constraints, 77, 132-33n16, 133n17
automata, 81
automation, 81-82
Auto-Sembly process, 64

autosuggestions, 83
Ayrton, William Edward, 12, 113n42

Bartlett, J., 56
Becquerel, Edmond, 32, 121n47
belief, 4, 23, 34, 39, 97
Benjamin, Ruha, 79–80
Benjamin, Walter, 46, 61
big data, xi, 114–15n50, 115n55
Bishop, Leon W., 113n44
black box, 1, 4, 65, 87, 88, 89
Black digital practice, x
Bloch, Marc, 19–25, 28–32, 37–39; executed by Nazis, 39; *The Royal Touch*, 23; "Transformations of Techniques as a Problem of Collective Psychology," 28–30; World War I experiences, 24–25; Works: *French Rural History*, 25, 30
Blum, Hester, 52
body: as first instrument, 25, 33, 118n26; manual expertise of, 16, 23–27, 32–33, 35, 68, 74–75. *See also* body techniques; individual
body techniques, x, xii, 25–28, 31, 37
boredom, 77–78
boundaries, 9–11, 77, 132–33n16
Bradley, Eleanor Spencer, 56
Bratton, Benjamin, 82–83
Braudel, Fernand, 117n13, 120n38
Bredekamp, Horst, 31
Brock, David C., 138n3
Brown, Robert, 48–50, 125–26n17
Bucher, Taina, 34
Byrum, Greta, 72

cable crimping, 71
Cahiers Politiques, 39
Calahan, Harold, 67, 130n64
calendars, 40

capitalism, 89; surveillance, 75–76, 78
care, relationships of, 73, 132n10
carriage knife, 68
catchpiece, 59
Center for Humane Technology, 76
Certeau, Michel de, 89
chaîne opératoire, 10–11, 111n35, 112–13n40, 113n41
Chambers, Frank, 107n6
ChatGPT, 61, 96, 99
Choreography for Smartphone Gestures (Ramírez), viii (fig.)
Chun, Wendy, 119n34
cinema, 25, 109n14
Clarke, Arthur C., 33–34, 97, 108n8
colonialism, and data extraction, 114–15n50
community technology, 73–75, 78, 91–92
comparativism, 29, 37, 40–41
conspiracy theories, 6
constraints, cultural, 77, 132–33n16, 133n17
consumer electronics, 64–65
contrôle technique, 20–21
coterie speech, 51, 52
Coupland, Douglas, 80–81
Covid-19 pandemic, 6, 8, 56, 72–74, 131n4; "Great Pause," 109n19
Cowan, Ruth Schwartz, 89
Crary, Jonathan, 27, 105n10, 119n32
Crawford, Matthew, 77, 133n17
cryptocurrency, 92
cultural techniques (*Kulturtechniken*), x–xi, 31, 115n52
Cultural Techniques (Siegert), 15, 30–31
culture, 16; cultural imperatives, 97, 107n3, 107n6; historical semantics of, 31; outpaced by technical systems, 39–40

cyborg consciousness, 65
"Cyborg Manifesto" (Haraway), 10–11

Dana, Richard Henry, Jr., 52–53
Darwin, Charles, 40, 122n59
data brokers, xiii, 14
Daumas, Maurice, 118n20
death, transcendence of, 4–5, 97, 140n13
Delany, Samuel R., 65, 86
Detroit Community Technology Project, 73
Devonshire Association, 44
dictionaries, 51, 54–60
Dictionary (Johnson), 60
digital minimalism, 75–80, 85, 134n26; and elitism, 78–80; medicalized discourse, 76–77
dispositif, 59, 105n4
distributional hypothesis, 57
Durkheim, Émile, 21, 24, 32, 36, 116n3, 121n47

Edgerton, David, 112n39
Edison, Thomas, 4
Edwards, Mary K. Bercaw, 50, 51, 125n14
Elements of a Philosophy of Technology (Kapp), 36
elitism, and digital minimalism, 78–80
Ellul, Jacques, 40
embodied intelligence and know-how, 7, 9, 15, 88, 136–37n48; in post-smartphone era, 91
ensembles, technical, 11, 32–33, 38, 80
epistēmē, x
equity, digital, 6, 17, 71–74, 131n4, 131n7, 132n12; and digital minimalism, 79–80
Escape from New York (movie), 6
Espinas, Alfred, 36, 122–23n62, 123n64, 123n5, 135n37
ether, 4, 108–9n12
Ethernet cables, 71, 92
etymology, 55–56
etymon, 56
everyday practices of users, xi–xii, 8–10, 112n36; automation of, 75, 89; extraction of by tech business model, 14; influenced by social media algorithms, 34. *See also* users
expertise, xiii, 15–17, 136n42; emergence of new epistemologies from, 85; manual, 16, 23–27, 32–33, 35, 68, 74–75
extraction of value, 14, 98, 100; data extraction, 114–15n50, 115n55

Febvre, Lucien, 19, 21, 24
Figuier, Louis, 32, 121n47
film theory, 59
Firth, John, 57, 127n39, 128n42
folk theories, 8, 34
formal specifications, 87
Fossey, Charles, 20
Foucault, Michel, x, 105n1
France: and agricultural revolution, 28–29; *contrôle technique*, 20–21; French resistance, 39; Toulouse conference of 1941, 16, 19–21, 23–24, 28, 31–32, 38, 116n1, 117n13; Vichy government, 19–21, 96, 116n2, 116n5; wartime food rationing, 32
French Rural History (Bloch), 25, 30
Frosh, Paul, 10

INDEX

gâche (part of a latch), 59
gâchette (hinge, toggle, trigger), 58–59
gadget, 16–17, 43–69; cheaply made tools implied by, 62; in context, 58–59; cumulative history of application, 63–64; evolving meanings of, 46–47, 51–54, 58, 62–64; French correlatives to, 58–59; as functional and fictional device, 53–54; ideas indicated by, 62–63; indeterminacy of, 61–62; as nostalgic, 50; in *OED*, 58; semantic drift of meaning, 47, 56, 62–65; sensory world of, 58; in spoken language of mid-nineteenth-century sailors, 51; tool, use, and description, categories of, 66–68
Gadgets and Wrinkles (Calahan), 67
gaget, 58
Gaget, Gauthier, & Co., 58
General Commissariat on Jewish Questions, 21
Gille, Bertrand, 39–40, 118n20
glamour, 60
Google Paper Phone, 76
grammar, 60
Gramsci, Antonio, 110n27
Great War era, 1–5
Greenfield, Adam, 82
Guillory, John, 111n30

habit, x, xii, 6, 7, 10, 12; automatism, 85; and body techniques, 27–28; framed as digital addiction, 76, 78, 80, 119n34
Hair in the Cable, The (Ọnụọha), 74
Hall, Stuart, 110n28
ham radio sets, 64

hand, as tool, 36, 68
Haraway, Donna, 10–11, 65
Harris, Tristan, 76
Hendren, Sara, 118n26
hinge, 59, 69
historical change, and techniques, 22–23
historiography, 117–18n20; Annales school, 21, 23–24, 27
History of Techniques (Gille), 39–40
Hobart, Hiʻilei Julia Kawehipuaakahaopulani, 132n10
hope, 96–97
How to Do Nothing (Odell), 79
Hubert, Henri, 32
humanity, techniques to prove, 99–100
Hurm, Horace, xiv (fig.), 107n6

iCloud, 7
Igbo culture, 74
individual: agency of, 18; individuation, 40; scaled up to scale of civilizations, 31–32. *See also* body
industrialization, 35, 38
influenza pandemic of 1919, 1–6, 108n10, 108n11
Ingold, Tim, 15–16, 113n41
inscrutability, 61, 75, 137n50; fallacy of, 88–89
instruction manuals, 68–69
internet: and digital equity, 6, 17, 71–74, 131n4, 131n7, 132n12; manuscript and print metaphors for, 50; nautical terminology for, 50–51, 66, 126n22, 126n23
intranets, 100
invention, 112n38; from bottom-up, 29–30; "societies of," 28–29
iPhone, 7

iPod, 12
IRL Glasses, 76

James, William, 85
Jen, Gish, 83-84, 96
Jobs, Steve, 7
Johns, Adrian, 51
Johnson, Samuel, 60

Kapp, Ernst, 36, 41
Keller, Charles, 113n41
Kinematics of Machinery (Reuleaux), 36
Kipling, Rudyard, 45
Kirschenbaum, Matthew, 99
Kittler, Friedrich, 4, 109n14
Kneese, Tamara, 132n10
knowledge, 7-8; embodied know-how, 9, 15, 136-37n48; tacit, 8, 61, 88-89; using disassociated from knowing, 86-87, 136n42
Kracauer, Siegfried, 77-78
Krämer, Sybille, 31
Kroll, Joshua, 89, 137n50
Kulturtechniken (cultural techniques), x-xi, 31, 115n51

La Boulangerie, 24
La charrue (Legros), 18 (fig.)
la langue and *la parole*, 99
League of Free Men (Bund freier Männer), 36
Lefebvre des Noëttes, Richard, 30, 120n38
Legros, Alphonse, 18
Leroi-Gourhan, Andre, 11, 25, 39, 136n47
Les Merveilles de l'Indus (Figuier and Becquerel), 32
Les Origines de la Technologie (Espinas), 36

Lévi-Strauss, Claude, 39
lexicography, 127n33, 127n39; NLP compared with, 57; textual evidence, 51-52; women's participations in, 55-57
Liberman, Anatoly, 55
Light Phone, 76
listening, techniques of, 27-28, 114n45, 119n33
Liu, Lydia H., 57, 128n42
Lodge, Oliver, 4
Luddism, 6, 79, 133n22

Ma, Ling, 5
MacFarlane, William (MacFarlanes), 1-5, 11, 12
machine learning, ix, 61, 87, 95, 138n1
magic, 23, 32-35, 97, 117n15; cat's cradle trick, 60-61
making do, 16, 115-16n56
Manual of Seamanship, 42 (fig.)
manuscript and print technologies, metaphors from, 50
Marconi, Guglielmo, 11-12, 48
Marine Engineering, 45
marline, marling, 47, 60, 66
marlinspikes, 50, 53, 66-67
marl or *marl down*, 60, 66
Marvin, Carolyn, 10
Marx, Karl, 105n10, 119-20n37
Massachusetts Institute of Technology (MIT), 13-14
Masterman, Margaret, 57, 127n40
Mauss, Marcel, x, 9, 19-28, 80, 100; arrest and execution of, 39, 117n8; ethnography course, 32, 120n44; and La Boulangerie, 24; magic, view of, 23, 33-35, 97, 117n15; program for study of technique, 36-38, 121n47;

technical ensembles, concept of, 32–33, 38, 80; "Techniques and Technology," 31–33, 39, 121n49; "Techniques of the Body," 25–26, 31; utopian hope of, 96–97; World War I service, 24, 116n3
Mavhunga, Clapperton, 12, 16, 115n55
McLuhan, Marshall, 63
mechanical discourses, 35
media, lack of fixed identities, 10–11
media ecology, 10, 112n36
media theory, 10, 27; Anglophone, 40; from below, 9; comparativism, 29, 37, 40–41; French, 39; German, 15, 30–31, 40
medieval world, technical systems in, 23
medium, 9–10, 108–9n12, 111n31
Melville, Hermann, 59–60, 68, 125–26n17
Merchant, Brian, 10
mesh nodes, 71–73, 91–93
Meyerson, Ignace, 19, 20, 21, 24, 39, 116n1
military industries, 16, 125n12
Mills, Mara, 27–28, 119n33
misinformation, 5, 6, 8, 75
Moby-Dick (Melville), 59–60, 68
Moi, Toril, 47
Morse code, 7

Nadar, Félix, 97
natural language processing (NLP), 57–58, 61, 128n42
natural selection, 40, 122n59
nautical terminology, 57–58; for internet technologies, 50–51, 66, 126n22, 126n23
neoliberalism, 132n10, 132n12, 133n19
neology and novums, 53
New School, 72
nineteenth-century thinkers, 35–36
NoPhones, 76
Notes and Queries, 43, 44, 45, 52, 54
Nova, Nicolas, 34
novelty, used to outsmart AI, 99–100
Nucera, Diana, 73

objects, imagined historical timeline for, 35–36
Odell, Jenny, 79
"offline is the new luxury," 79–80
Oldenziel, Ruth, 89
Ondophone, xiv
On the Mode of Existence of Technical Objects (Simondon), xii, 40
Onụoha, Mimi, 74
organon, 36
Oudshoorn, Nelly, 89
Oxford English Dictionary (OED), 54–58

pattern of expectation, 3, 108n8
Penn, Michael, 70 (fig.)
personas (fictional users), 82–83
Pétain, Philippe, 19, 116n2
phenakistoscope, 27
Philadelphia, 3, 17
Philly Community Wireless (PCW), 71–74, 91–93, 131n3; mesh nodes, 71–73, 91–92, 131n3
Philological Society, 54–55
philosopher-technicians, 36, 41, 47
Philosophy of Manufacturers, The (Ure), 35
phone, prefixes, 66
photography, 97, 140n11

Pinch, Trevor, 89
pirates, 51, 126n23
Pitt-Rivers, Augustus, 35–36, 122n59
"poaching"/tactics, 89–90
pocket wireless, 1–6, 11–12, 97, 107n3, 107n6, 113–14n44
Polanyi, Michael, 88
politics, plow as foundational act of, 30–31
Portwood-Stacer, Laura, 78
practice, xiii, 24–25, 85, 123n62; communities of, 45, 51, 60–61, 74–75, 92; regions of, 47, 51; unconscious, 135n37
praxiology, 36, 123n62, 123n64
pretrial sentencing algorithms, 79
Psychical Society, 4

Race after Technology: Abolitionist Tools for the New Jim Code (Benjamin), 79–80
racial inequality, 79–80, 131n6
radio frequency waves, 1–4, 97; "electronic kinship," 5, 109n18; and hope for transcendence of death, 4–5, 97
Ramírez, Naomi Elena, viii
Raymond, or Life and Death (Lodge), 4
RCA logo (His Master's Voice), 97, 140n12
Redburn (Melville), 125–26n17
refusal, x, 78, 100
regulatory policy, lack of, 38, 133n19
religion, 32–33
resistance: to new agricultural techniques, 28–29; workplace, 133–34n23, 135n32
Resisters, The (Jen), 83–84, 96
Reuleaux, Franz, 36, 41
Risam, Roopika, 16

ritual, 23, 32–34; "conscious clicking," 34
Rose, Ellen, 86, 136n42
Royal Touch, The (Bloch), 23
Rozwadowski, Helen, 48
Russell, Lindsay Rose, 55

sabotage of wireless towers (2020), 6
Sackville-West, Vita, 61–62
sailors, 17, 47–54; knots, 9, 42 (fig), 50, 60, 67; literacy of, 52; nautical terminology, 50–51, 57–58, 66, 126n23; occupational lingo versus coterie speech, 51, 52; ropework, knowledge of, 49–50; "spinning a yarn," 48–49, 125n14; steamship technologies, 48, 62; wrinkles as procedure, 67, 130n65
Saussure, Ferdinand de, 99
Schatzberg, Eric, 15, 116n57
Schiffer, Michael B., 105n3
Schlanger, Nathan, 112–13n40
science, history of, 40
Scott (amateur lexicographer), 55, 57
scrolling, 26–27, 76, 85
semantic drift, 47, 56, 62–65
set, meanings of, 55
Severance (Ma), 5
Shotwell, Alexis, 80, 134n26
Siegert, Bernhard, 15, 30–31
"sign language" of tools, 65
Simiand, François, 29, 32
Simondon, Gilbert, xii, 40
Skipper, Ellen, 55, 57
slang, spoken, 51
slavery, 30
smartphones: blurring of aesthetic borders by, 9–11; and blurring of boundaries, 9–11, 77; fragility of, 90; as gadgets,

47; manual dexterity in use of, 26; operation at gigahertz level, 2; post-smartphone tool kit, 90–91; rituals associated with use, 34; scrolling, 26–27, 76, 85; as "the one device," 10; unconscious techniques, 87–88; user experience design, 68–69; users as addicted, 66, 75–76, 85–86, 119n34; in workplace, 79, 133–34n23

social media, x, 26, 34, 78, 85, 90

social sciences, 21

"societies of invention," 28–29

"societies of routine," 28–29

Sondheimer, Janet, 22

Sontag, Susan, 7

Sony Walkman, 12

Sparke, Archibald, 43–44

"spinning a yarn," 48–49, 125n14

spiritual beliefs and occult practices, 4–5, 97, 108–9n12

Spunyarn and Spindrift: A Sailor Boy's Log of a Voyage Out and Home in a China Tea-Clipper (Brown), 48–50, 53, 54, 58, 125–26n17

Stanley, Jason, 136–37n48

Statue of Liberty souvenirs, 58

Statut des Juifs (France), 19–20

steamship technologies, 48, 62

stereoscope, 27

Sterne, Jonathan, 27–28, 119n33, 140n14

Subramanian, Samanth, 95

surplus data, 115n49

surveillance, 8; capitalism, 75–76, 78; of habit, 14, 17, 69; of on-the-job performance, 82, 135n32

tacit experience, 8–9, 13, 15

tacit knowledge, 7, 88

Taggart, John, 33, 121n50

Tallinn, Jaan, 138n3

Target, 82

tech, 114n48; as business model, 14; high-stakes race with technique, 99–100

technical facts, 30, 31

technical systems, cultural development outpaced by, 39–40

technicians, 16–17, 40–41, 86; in discipline of technology, 17, 37, 121n47; philosopher-technicians, 36, 41, 47. *See also* sailors; users

technics, xi; as academic field, 12–13; technique differentiated from, xii–xiii

Technik, x–xi, 31, 115n52

technique: beyond Western origins, 16, 115n55; cognates in other languages, 15–16; combined categories, 35; as common language, xi–xiii; connotations of, 14–15; conscientious, 74–75, 87, 91–92; conscious and unconscious, 87–88, 135n37; as fossil fuel of generative AI, 98; French meanings of, 22–23; high-stakes race with tech, 99–100; and historical change, 22–23; individual scaled up to scale of civilizations, 31–32; of listening, 27–28; magic's resemblance to, 32–33; micro-level, 8–9; as new object of study, 39; and planning, 38; precision lacking in term, 24; reclaiming, 95–100; and social tradition, 23,

INDEX

30, 37; tool as, 15, 22. *See also* body techniques
"Techniques and Technology" (Mauss), 31–33, 39, 121n49
"Techniques of the Body" (Mauss), 25–26, 31
Techniques of the Observer (Crary), 27
technobabble, 53
technocratic worldview, 84–85
technologie, 36–40; as academic discipline, 32, 40; as "science dealing with techniques," 36–37; sources, 37; as study of techniques, 36–37, 47; versus *Technologie*, 36; workers and technicians in discipline of, 37, 40
technology: ages of, 9; changing definitions of, 13; monocausal explanations, 29–30; parasitic relationship with, 86, 136n42
"Techno-optimist Manifesto," 98, 140n16
tékhnē, x–xii, 87; Aristotelian conception, 15–16
telegraph, 11–12, 48
terminology, evolving meanings of, 13, 46–47, 64
textpocalypse, 99
theories of technology: high and low, 9, 17
thesauri, 57
Third Law (Clarke), 33–34
toggle, 59, 67, 69
tokenization, stemming, and lemmatization, 56–57
tools: as form of "organ projection," 36; gadgets, 16–17; as metaphor for words, 46, 50; as technique, 15, 22; as unknowable, 7, 87–89, 138n3
Toulouse conference of 1941, 16, 19–21, 23–24, 28, 31–32, 38, 116n1, 117n14
Traffics and Discoveries (Kipling), 45
"Transformations of Techniques as a Problem of Collective Psychology" (Bloch), 28–30
Tuhus-Dubrow, Rebecca, 12
Tune In (Penn), 70 (fig.)
Turner, A. M., 56
Two Years before the Mast (Dana), 52–53

United Nations, 73
United States, German exiles in, 36
University of Minnesota Press, 139n7
Ure, Andrew, 35
useful arts, 12, 35
user experience design, 68–69
users: as addicted, 66, 75–76, 85–86, 119n34; agency of, 9, 17, 89; common man replaced by, 89; early days of computing, 86; generations of, 66; nonspecialist and skilled artisans, 85–86; normalization of mad, fantastic, or science fictional techniques, 12; unique syntax of, 12; using disassociated from knowing, 86–87, 136n42. *See also* everyday practices of users; technicians
using, disassociated from knowing, 86, 136n42

Vichy government, 19–21, 96, 116n2, 116n5
vision, "historical construction" of, 27, 119n32
Vismann, Cornelia, 31

Wajcman, Judy, 89
walking fashions (gaits), American, 25
Ward, Megan, 79
Webster, H. T., 94 (fig.)
whale, etymology of, 59–60
Whitehead, Alfred North, 136n45
Williams, James, 77, 132–33n16
Winchester, Simon, 56
Winthrop-Young, Geoffrey, 15, 115n52
wireless: pocket experimenters, 1–6, 11–12, 97, 107n3, 107n6, 113–14n44; sabotage, 2020, 6; between ships, 51, 125n14; telegraph, 11–12, 48
Wittgenstein, Ludwig, 46, 57, 128n42
women: as innovators, 89, 137n52; and lexicography, 55–57
words: social context of, 57–58; tool metaphor for, 46, 50. *See also* gadget
World War I, 1–5, 11, 19, 24–25, 44, 116n3
World War II, 16–17, 19–20, 22, 24, 64
wrinkles, 67, 130n65
Writers Guild of America, 96
writing, erosion of, 95

(continued from page ii)

37 *Noise Channels: Glitch and Error in Digital Culture*
Peter Krapp

36 *Gameplay Mode: War, Simulation, and Technoculture*
Patrick Crogan

35 *Digital Art and Meaning: Reading Kinetic Poetry, Text Machines, Mapping Art, and Interactive Installations*
Roberto Simanowski

34 *Vilém Flusser: An Introduction*
Anke Finger, Rainer Guldin, and Gustavo Bernardo

33 *Does Writing Have a Future?*
Vilém Flusser

32 *Into the Universe of Technical Images*
Vilém Flusser

31 *Hypertext and the Female Imaginary*
Jaishree K. Odin

30 *Screens: Viewing Media Installation Art*
Kate Mondloch

29 *Games of Empire: Global Capitalism and Video Games*
Nick Dyer-Witheford and Greig de Peuter

28 *Tactical Media*
Rita Raley

27 *Reticulations: Jean-Luc Nancy and the Networks of the Political*
Philip Armstrong

26 *Digital Baroque: New Media Art and Cinematic Folds*
Timothy Murray

25 *Ex-foliations: Reading Machines and the Upgrade Path*
Terry Harpold

24 *Digitize This Book! The Politics of New Media, or Why We Need Open Access Now*
Gary Hall

23 *Digitizing Race: Visual Cultures of the Internet*
Lisa Nakamura

22 *Small Tech: The Culture of Digital Tools*
Byron Hawk, David M. Rieder, and Ollie Oviedo, Editors

21 *The Exploit: A Theory of Networks*
Alexander R. Galloway and Eugene Thacker

20 *Database Aesthetics: Art in the Age of Information Overflow*
Victoria Vesna, Editor

19 *Cyberspaces of Everyday Life*
Mark Nunes

18 *Gaming: Essays on Algorithmic Culture*
Alexander R. Galloway

17 *Avatars of Story*
Marie-Laure Ryan

16 *Wireless Writing in the Age of Marconi*
Timothy C. Campbell

15 *Electronic Monuments*
Gregory L. Ulmer

14 *Lara Croft: Cyber Heroine*
Astrid Deuber-Mankowsky

13 *The Souls of Cyberfolk: Posthumanism as Vernacular Theory*
Thomas Foster

12 *Déjà Vu: Aberrations of Cultural Memory*
Peter Krapp

11 *Biomedia*
Eugene Thacker

10 *Avatar Bodies: A Tantra for Posthumanism*
Ann Weinstone

9 *Connected, or What It Means to Live in the Network Society*
Steven Shaviro

8 *Cognitive Fictions*
Joseph Tabbi

7 *Cybering Democracy: Public Space and the Internet*
Diana Saco

6 *Writings*
Vilém Flusser

5 *Bodies in Technology*
Don Ihde

4 *Cyberculture*
Pierre Lévy

3 *What's the Matter with the Internet?*
Mark Poster

2 *High Technē: Art and Technology from the Machine Aesthetic to the Posthuman*
R. L. Rutsky

1 *Digital Sensations: Space, Identity, and Embodiment in Virtual Reality*
Ken Hillis

GRANT WYTHOFF directs graduate student programs at the Center for Digital Humanities, Princeton University. He is cofounder of the cooperative mesh network and digital equity organization Philly Community Wireless and is editor of *The Perversity of Things: Hugo Gernsback on Media, Tinkering, and Scientifiction* (Minnesota, 2016).